Fundamentals of Natural Gas

Fundamentals of Natural Gas

Editor

Raju Danwate

Fundamentals of Natural Gas

Edited by **Raju Danwate**

Printed in 2017

ISBN: 978-1-68117-380-1

Library of Congress Control Number: 2015936530

© 2016 by
SCITUS Academics LLC,
616, Corporate Way, Suite 2, 4766,
Valley Cottage, NY 10989

www.scitusacademics.com

Contents

Preface

Natural gas has been a valuable energy commodity for many centuries. Although natural gas was unpopular prior to the eighteenth century due to the use of manufactured gas such as 'coal gas', it was towards the latter part of the nineteenth century that most industrial countries started using natural gas and thus large transmission and distribution pipelines were constructed in transferring the gas to the required areas. The steady utilisation of natural gas grew to the peak during the 1960s to mid-1970s when the shortage of crude oil enforced most industrial nations to find alternative ways of harnessing energy and natural gas has since become one of the main fossil fuel energy sources. Natural gas is colourless with high flammability and energy value and together with its convenience has resulted in a rapid rise to extensive use as a fuel today.

Editor

Effect of Temperature Shock and Inventory Surprises on Natural Gas and Heating Oil Futures Returns

John Wei-Shan Hu[1, 2], Yi-Chung Hu[1], and Chien-Yu Lin[3]

[1]Department of Business Administration, Chung Yuan Christian University, Chung Li 32023, Taiwan

[2]Department of Finance, Chung Yuan Christian University, Chung Li 32023, Taiwan

[3]Huang Lin Co., Taoyuan 33742, Taiwan

ABSTRACT

The aim of this paper is to examine the impact of temperature shock on both near-month and far-month natural gas and heating oil futures returns by extending the weather and storage models of the previous study. Several notable findings from the empirical studies are presented. First, the expected temperature shock significantly and positively affects both the near-month and far-month natural gas and heating oil futures returns. Next, significant temperature shock has effect on both the conditional mean and volatility of natural gas and heating oil prices. The results indicate that expected inventory surprises significantly and negatively affects the far-month natural gas futures returns. Moreover, volatility of natural gas futures returns is higher on Thursdays and that of near-month heating oil futures returns is higher on Wednesdays than other days. Finally, it is found that storage announcement for natural gas significantly affects near-month and far-month natural gas futures returns. Furthermore, both natural gas and heating oil futures returns are affected more by the weighted average temperature reported by multiple weather reporting stations than that reported by a single weather reporting station.

INTRODUCTION

During the past four decades, energy consumption has fluctuated markedly owing to fluctuation in energy demand and supply, as well as significant changes in climate conditions. Along with climate change and an increase in disasters frequency, global warming is a serious concern worldwide. Ruth et al. [1] argued that climate change, with associated events ranging from rising sea levels to strong and frequent storms and extreme temperature events, will significantly impact the natural environment and human infrastructure and its contribution to economic activity and quality of life. These impacts increase direct and indirect costs accrued from increasing environmental damage and disruption.

According to Chicago Mercantile Exchange (CME), the weather directly affects nearly 30% of the US economy. The US Energy Department estimated that $1 trillion of the US economy was exposed to weather risk in 2011. However, the notional value of traded weather derivatives was around US $3.5 billion, representing a mere fraction of total exposure. Consequently, weather derivatives are financial instruments provided by organizations or individuals to reduce or transfer risk associated with adverse or unexpected weather events. Organizations or individuals quantify weather in terms of how much

Effect of Temperature Shock and Inventory Surprises on Natural Gas and...

3

temperature, frost, hurricane damage, or snowfall deviates from the monthly or seasonal average in a particular city or region. However, it is estimated that approximately 98.0% of currently traded weather derivatives are based on temperature. The first weather derivatives transaction was executed in the summer of 1997 by Aquila Energy as a weather option embedded in a power contract [2]. Gas, oil, and power companies use heating degree days (HDD) or cooling degree days (CDD) contracts to smooth earnings. HDD and CDD are among the most common weather derivatives.

This study examines and compares the impact of the temperature shock, expected natural gas and heating oil inventory surprises, movement of the Dow Jones industrial index (DJ), winter effect, storage announcement effect, hurricane announcement effect, and the nonlinear temperature effect on the conditional means and volatility of both near-month and far-month natural gas and heating oil futures returns.

LITERATURE REVIEW

Previous literatures are classified into two categories: one on the impact of temperature or weather on commodity futures; the other on natural gas and/or heating oil futures. Relevant studies in the first category include the following: Stevens [3] pointed out that the weather and climatology literature indicated persistence in North American weather patterns during the summer months. Given this nonrandom character of weather and given that the corn, wheat, and soybean belts are sufficiently geographically concentrated to be dominated by a regional weather phenomenon, their futures markets are hypothesized to reflect this assimilation of nonrandom weather information as nonrandom price fluctuations. Ates and Wang [4] found that extreme cold weather and inventory surprises influenced variation in basis, spot, and futures price changes. Furthermore, the conditional volatility of natural gas and heating oil spots and futures markets was higher in winter and lower in summer. Mu [5] examined how weather shocks influenced asset price dynamics in the US natural gas futures market. The empirical results revealed a significant weather effect on both the conditional mean and volatility of natural gas futures returns. Combined with the evidence that the volatility is significantly higher on Monday and on the day when the natural gas storage report is released, their findings suggested that information on market fundamentals significantly determines natural gas volatility. Chen et al. [6] examined the role of weather as a short-term demand factor and inventory as a short-term supply factor in explaining price spikes and time-varying volatility in natural gas spot and futures returns.

For studies in the second category, Herbert [7] summarized the relationship between spot and futures prices for natural gas, which could obtain accurate forecasts of spot prices. The natural gas futures market, thus, appeared inefficient. Walls [8] pointed out that the natural gas futures market was generally consistent with the efficient market hypothesis; that is, the futures market price was an unbiased predictor of spot prices at most market locations examined. Chinn et al. [9] examined the relationship between spot and futures prices for energy commodities (crude oil, gasoline, heating oil, and natural gas). Chinn et al. found that futures accurately predicted future spot prices, with the exception of 3-month natural gas futures. Chiou-Wei et al. [10] identified empirical regularities between changes in futures prices and surprise changes in natural gas in storage. Chiou-Wei et al. found an inverse relation between changes in futures prices and surprises in the change in natural gas in storage. Suenaga et al. [11] found that the volatility dynamics of NYMEX gas futures displayed two important features: (1) volatility is greater in winter than summer and (2) the persistence of price shocks and, hence, the correlation among currently traded contracts exhibited considerable seasonal and cross-sectional variation, consistent with the theory of storage. Recently, the usefulness of computational intelligence tools is highlighted by applying related models, such as neural networks and fuzzy sets, to natural gas consumption [12–14].

METHODOLOGIES

This study examines the influence of temperature change shock and inventory surprises on returns of near-month and far-month natural gas and heating oil futures returns from 2003 to 2006, extending the expected temperature and natural gas inventory shock model presented by Mu [5] to examine both natural gas and heating oil futures, and uses the Dow Jones index (DJ) provided by DataStream to proxy for equity market return. This investigation uses the daily temperature data from January 1, 2003 to December 31, 2006 provided by the NYMEX. Hurricane daily data is obtained from network of NYMEX and EQECAT. The other data sources are obtained from US Energy Information Administration and National Oceanic and Atmospheric Administration (NOAA). This investigation also examines whether the temperature reported by a single weather reporting station or the weighted average temperature reported by multiple weather reporting stations is more appropriate for examining the impact of temperature shock on energy futures returns. This study includes DJ returns, winter heating season, energy announcements, hurricane announcements, and nonlinear temperature as the independent parameters. As defined by contract 1 (namely, near-month) and contract 2 (namely, far-month) illustrated by EIA, contract 1 is a futures contract specifying the earliest delivery date; meanwhile, contract 2 represents the subsequent delivery month to that in contract 1.

To achieve the above objectives, the parameters are defined on natural gas and heating oil futures first; then the ADF test is applied to examine whether the parameters of the relevant models are stationary. Upon handling the stationary problem, this work examines whether the model is characterized by self-autocorrelations. If self-autocorrelation exists, the ARMA method is used to solve the self-autocorrelation problem of residuals. Furthermore, this study employs Ljung-Box Q^2 and ARCH-LM methods to test for the ARCH effect. If the answer is positive, the GARCH model is used; if it is negative, then the ordinary least square (OLS) model is used.

Parameter Definition

Rate of Return

The rate of return (ROR) on energy futures is calculated as follows:

$$RET_{Z,i,t} = \left(\ln Z_{i,t} - \ln Z_{i,t-1} \right) \times 100,$$

(1)

where Z represents natural gas closing price or heating oil closing price and $Z=N$ suggests natural gas, while $Z=H$ indicates heating oil; furthermore, $RET_{z,i,t}$ denotes natural gas or heating oil futures returns at day t; $i=1$ suggests nearmonth futures, while $i=2$ demonstrates far-month futures.

$\ln Z_{i,}$ and $\ln Z_{i,t-1}$ represent the nature log of the closing price of energy futures at day t and day $t-1$, respectively.

Expected Temperature Shock

The expected temperature shock is calculated on the basis of the models presented in [5]. However, this work uses the weather derivatives products traded on NYMEX. This investigation then divides the temperature of the weather reporting stations into the temperature reports for a single weather reporting station and the weighted average temperature for multiple weather reporting stations which are calculated from the four largest energy consumption states. The weights are as follows. (1) For natural gas: California, 0.28; New York, 0.27; Illinois, 0.25; and Michigan, 0.20; (2) for heating oil: New York, 0.30; Pennsylvania, 0.28; New Jersey, 0.23; and Massachusetts, 0.19. Expected temperature change shock for natural gas and heating oil is calculated as follows:

$$DD_{z,j,t} = CDD_{z,j,t} + HDD_{z,j,t}$$

$$W_{z,j,t} = \frac{1}{n} \sum_{k=1}^{n} \left(DD_{z,j,t+k} - ADDN_{z,j,t+k} \right),$$

$$(2)$$

where $DD_{z,j,t}$ denotes the sum of the cooling degree days (CDD) and heating degree days (HDD) of natural gas; $j=1$ demonstrates the temperature reported by a single weather reporting station; $j=2$ suggests the weighted average temperature reported by multiple weather reporting stations. $CDD_{z,j,t}$ demonstrates the CDD of natural gas at day t, which is the weighted average temperature reported by four weather reporting stations. Furthermore, $HDD_{z,j,t}$ is the HDD of energy futures at day t; $W_{z,j,t}$ denotes the expected average temperature shock of energy futures at day t; $DD_{z,j,t+k}$ is the sum of the CDD and the HDD for energy futures at day $t+k$; $ADDN_{z,j,t+k}$ represents the average degree days of energy futures at day $t+k$ for the past 30 years; $n=7$ denotes the weather forecast for change in weather conditions for the next 7 days.

Expected Inventory Surprises

The change in expected inventory surprises for energy futures is calculated as follows:

$$E\left(\Delta I_{z,t}\right) = c_0 + c_1 TZ_t + c_2 TZ_t^2$$

$$+ \left[\lambda_l \sin\left(2\pi\frac{w(t)}{52}\right) + \theta_l \cos\left(2\pi\frac{w(t)}{52}\right)\right] + \mu_t, \tag{3}$$

Where

$$\mu_t = \rho_1\mu_{t-1} + \eta_t + \eta_{t-1}, \eta_t \sim N\left(0, 1\right), \tag{4}$$

where $E(\Delta I_{z,t})$ denotes the change in market expectations regarding inventory surprises for energy futures from the Friday of week $t-1$ to the Friday of week t; TZ_t represents the weighted weekly average temperature in week t of the energy futures; $w(t)$ is a repeating step function in the Fourier series that cycles through 1,2,...52 (namely, each week of the year). Based on Schwarz information criterion (SIC), the number of lags is set in the Fourier and autoregressive series to test for a serial correlation. The change in expected inventory surprise for energy futures is defined as the difference between the announced storage change and the expected inventory change:

$$EINVZ_\tau = \Delta I_{Z,\tau} - E\left(\Delta I_{Z,\tau}\right)$$

$$EINVZ_t = EINVZ_{\tau-1}$$

$$\text{when } IDZ_t = 1, \quad IDZ_t = 0, \text{ otherwise,} \tag{5}$$

Where $EINVZ_\tau$ and $EINVZ_{\tau-1}$ denote the forecasting error of the weekly inventory for energy futures at week's τ and $\tau-1$, respectively and extend weekly data into daily data. IDZ_t is a dummy variable of announced storage for energy futures. Since the weekly natural gas storage report is released by the EIA everyThursday, $IDN_t = 1$ for each Thursday, and $IDN_t = 0$ otherwise. On the other hand, the weekly heating oil reports released by the EIA use $IDH_t = 1$ for eachWednesday, and $IDH_t = 0$, otherwise.

Dow Jones Industrial Index Return

We calculate the Dow Jones industrial index (DJ) returns as follows:

$$DJ_t = (\ln DJ_t - \ln DJ_{t-1}) \times 100,$$

(6)

Where DJ_t denotes Dow Jones industrial index returns as day t. Ln DJ_t and ln $DJ_t{-}1$ represent the natural log of the closing prices of DJ on day's t and $t{-}1$.

Winter Heating Season

We use the definition of EIA, with the winter heating season WIN_t running from October to March. The month lies between October and March if $WIN_t = 1$, whereas $WIN_t = 0$, otherwise.

Hurricane Announcements (STM)

This investigation uses STM_t to represent hurricane announcements; the hurricane daily data[1] is obtained from NYMEX. The sample period runs from Jan. 1, 2003 to Dec. 31, 2006. $STM_t = 1$ denotes that a hurricane's degree is 3 or higher. Otherwise, STM_t is equal to zero.

Considered Models

The following models are taken into account: (1) augmented Dickey-Fuller (ADF) test; (2) Ljung-Box test [15];(3) autoregressive conditional heteroskedasticity (ARCH) test [16];(4) ordinary least squares (OLS) test or generalized ARCH (GARCH) test. First, we use the augmented Dickey-Fuller (ADF) test [17, 18] to examine whether the series data is stationary. The error term in the Dickey-Fuller test is autocorrelated or the set of time series models is complicated. The Ljung-Box test is then used to examine whether any of the groups of autocorrelations of a time series are different from zero. If the Ljung-Box test shows that most parameters do not have autocorrelation problem, the autoregressive moving average (ARMA) model is not required. Gujarati [19] pointed out that, despite the large sample, both the Box-Pierce Q and the LB statistics follow the chi-square distribution. However, the LB statistic has better small-sample properties than the Q statistic. Third, the ARCH model is used to characterize and model observed time series. The key idea of the ARCH model is that the variance of the current error term depends on the actual size of the squared error term of the previous time.

The Lagrange multiplier test [20] is employed to test lag length for ARCH errors, namely, ARCH-LM test. Fourth, the GARCH model, a generalization of the ARCH model, is used to examine the influence of temperature on energy futures returns. A typical GARCH (p, q) model (where p denotes the order of the GARCH terms and σ^2 and q represent the order of the ARCH terms) is proposed by Bollerslev [16] and specified as follows:

$$y_t \mid \Omega_t \sim N\left(x_t a, \sigma_t^2\right)$$

$$\varepsilon_t = y_t - x_t a,$$

$$\varepsilon_t \mid \Omega_{t-1} \sim N\left(0, \sigma_t\right)$$

$$\sigma_t^2 = \alpha_0 + \sum_{i=1}^{q} \alpha_i \varepsilon_{t-i}^2 + \sum_{j=1}^{p} \beta_j \sigma_{t-j}^2,$$

$$\tag{7}$$

Where x_t denotes the independent variable vector; a represents the vector of the regression coefficient; q refers to the lag time; $x_t a$ is the portfolio obtained from information set Ω_t; ε_t represents the error term between the actual and the estimated values; σ_t^2 represents the conditional variance of time series $1/t$. σ_{t-j}^2 Denotes the conditional variance of return RET time series at day $t-j$, where $j=1$ thru P.

The GARCH models is employed to test the mean and volatility of each studied parameter in relation to the influence of temperature on near-month and far-month natural gas and heating oil futures returns as follows:

$$\text{RET}_{Z,i,t} = a_0 + a_1 W_{Z,j,t} + a_2 \text{EINV}_{z,t} + a_3 D]_t + \varepsilon_t$$

$$\varepsilon_t \mid \Omega_{t-1} \sim N\left(0, \sigma_t^2\right)$$

$$\sigma_t^2 = \alpha_0 \alpha_1 \varepsilon_{t-1}^2 + \beta_1 \sigma_{t-1}^2 + b_1 \text{WIN}_t + b_2 \text{IDZ}_t$$

$$+ b_3 \text{STM}_t + b_4 W_{Z,j,t} + b_5 W_{Z,j,t}^2 ,$$

$$(8)$$

where Z represents natural gas or heating oil; $Z=N$ suggests natural gas; $Z=H$ indicates heating oil; $\text{RET}_{Z,i,t}$ denotes the ith month return of natural gas or heating oil futures at day t; $i=1$ suggests near-month futures; $i=2$ indicates far-month futures. Ω_t-1 denotes the total available information at day $t-1$; ε_{t-1}^2 demonstrates ε_{t-1}^2 the conditional variance is affected by the error term square at day $t-1$; $W_{Z,j,t} + w_z^2$ $'_{j,t}$ is the nonlinear effect of natural gas or heating oil at day t and jth temperature.

EMPIRICAL RESULTS

We examine the influence of temperature and inventory change shocks on natural gas and heating oil futures returns during 2003 to 2006. Upon deleting the missing data for any parameter, there are 988 original data samples for natural gas, and 991 for heating oil. Figure 1 depicts the returns of near-month and far-month natural gas futures. Figure 2 depicts the trend of temperature shock variable for natural gas for one weather reporting station (i.e., Chicago) and multiple weather reporting stations (i.e., California, New York, Illinois, and Michigan). Figure 3 shows the returns of heating oil for near-month and far-month futures. As for Figure4, it depicts the trends of temperature shock variables for heating oil for a single weather reporting station (i.e., New York) and multiple weather reporting stations (i.e., New York, Pennsylvania, New Jersey, and Massachusetts).

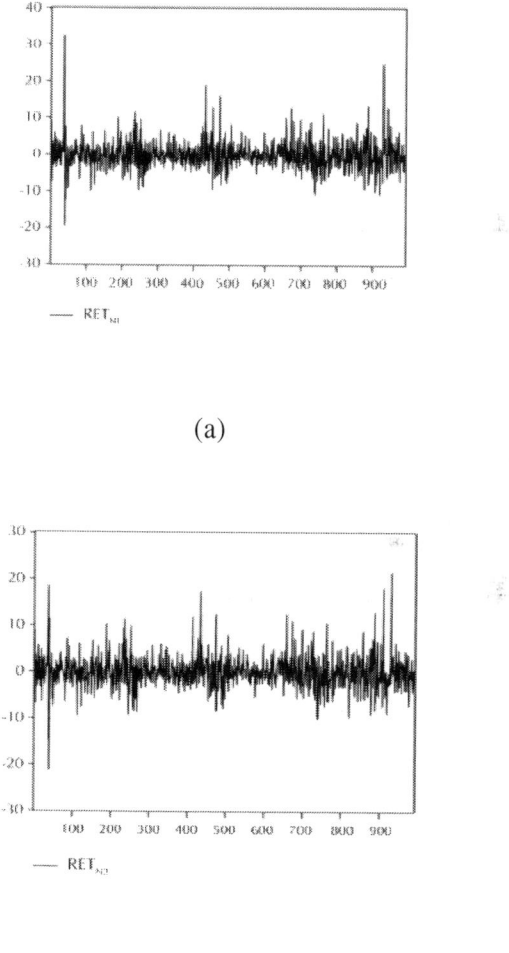

(a)

(b)

Figure 1: Return of natural gas: (a) near-month and (b) far-month.

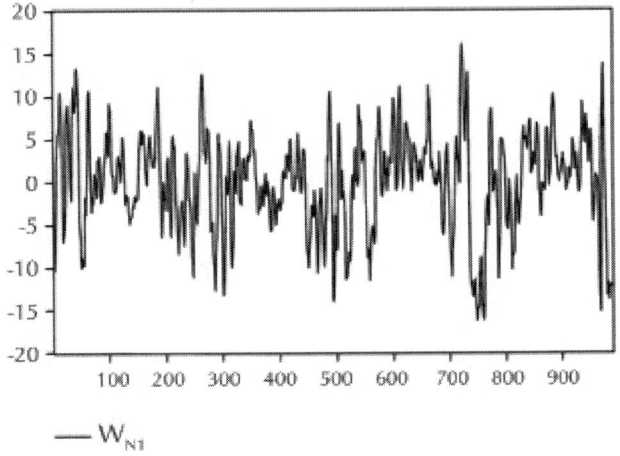

(a)

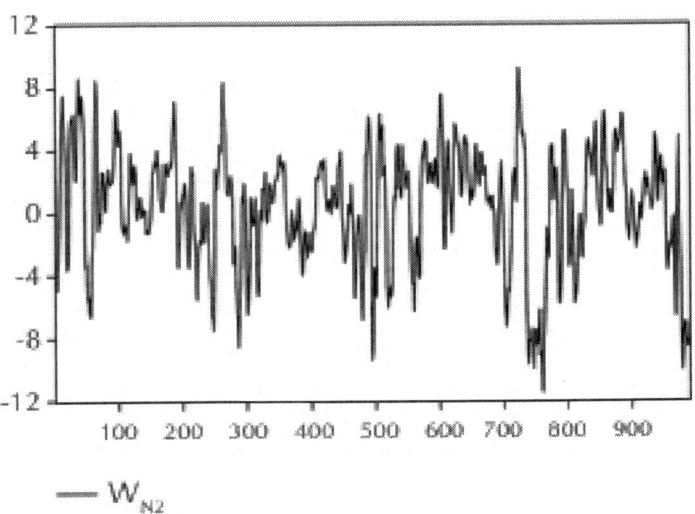

(b)

Figure 2: Temperature shock variable for natural gas.

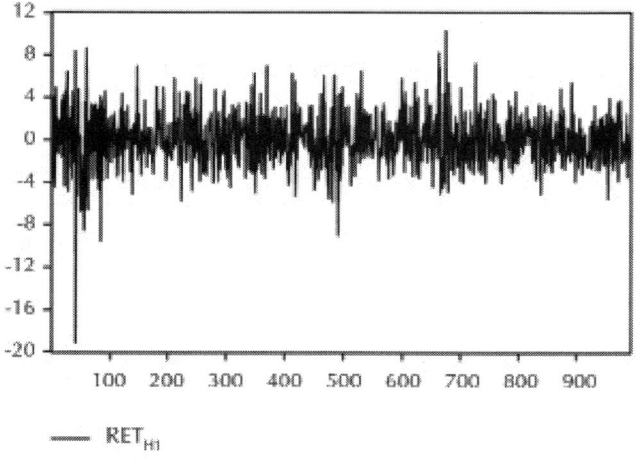

(a)

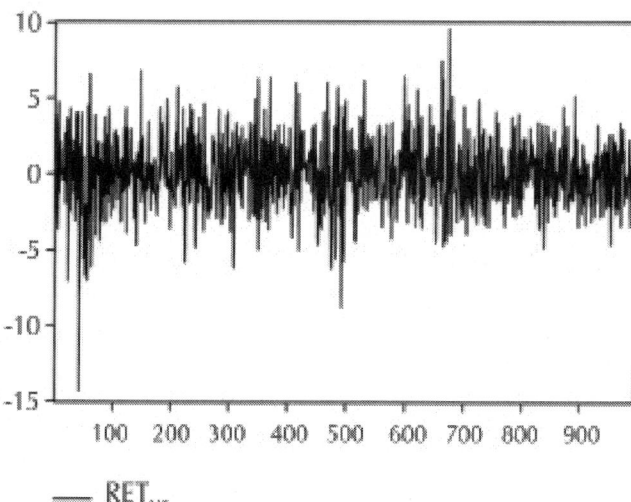

(b)

Figure 3: Returns of heating oil: (a) near-month and (b) far-month.

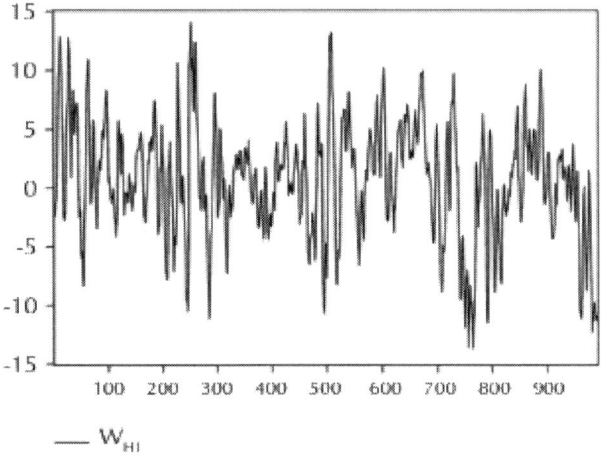

(a)

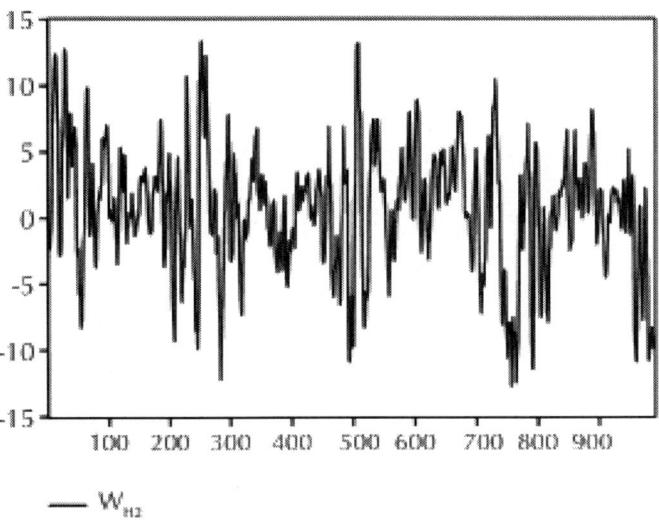

(b)

Figure 4: Temperature shock variable for heating oil.

Table 1 shows that each parameter in a time series sample is stationary. Ljung-Box Q tests are then employed to examine autocorrelation for all the natural gas and heating oil series. In Table 2, series 1 denotes energy futures returns are affected by the temperature reported by a single weather reporting, and series 2 denotes energy futures returns are affected by the weighted average temperature reported by four weather stations. (k) is the Ljung-Box statistic value of k-level time lag of returns series. Table 2 indicates that, under the Ljung-Box tests, most variables of series 1 and 2 for natural gas and heating oil futures are insignificant, suggesting that natural gas and heating oil futures are nearly free of autocorrelation. The autoregressive moving average (ARMA) model, thus, is not required.

Table 1: Results for ADF test

(a)

Natural gas model	ADF statistics-original series			PP statistics-original series		
Variables	Lag term	w/constant w/o time trend	w/constant & w/time trend	Lag term	w/constant w/o time trend	w/constant & w/time trend
RET 1	0	−31.7562***	−31.7437***	4	−31.7534***	−31.7411***
RET 2	0	−33.1140***	−33.1049***	1	−33.1135***	−33.1043***
1	6	−7.3796***	−7.3917***	37	−5.7894***	−5.8117***
2	9	−5.9611***	−6.0443***	21	−5.6500***	−5.7281***
EINVN	5	−6.0680***	−6.3519***	17	−7.4740***	−7.7667***
DJ	0	−33.7349***	−33.7173***	1	−33.7367***	−33.7191***

(b)

Heating oil model	ADF statistics-original series			PP statistics-original series		
Variables	Lag term	w/constant w/o time trend	w/constant & w/time trend	Lag term	w/constant w/o time trend	w/constant & w/time trend
RET 1	0	−33.4204***	−33.4068***	7	−33.4627***	−33.4494***

RET 2	0	−33.4655***	−33.4534***	9	−33.5213***	−33.5103***
1	6	−7.5365***	−7.8002***	26	−5.3673***	−5.5103***
2	6	−8.0559***	−8.3002***	26	−5.4189***	−5.5319***
EINVH	5	−6.5395***	−6.5298***	10	−7.1929***	−7.1846***
DJ	0	−33.7270***	−33.7095***	1	−33.7282***	−33.7107***

***1% significance level; **5% significance level; *10% significance level; RET_{N1}: natural gas near-month futures return; RET_{N2}: natural gas far-month futures return; RET_{H1}: heating oil near-month futures return; RET_{H2}: heating oil far-month futures return; W_{N1}: expected temperature shock for natural gas reported by single weather reporting station; W_{N2}: expected weighted average temperature shock for natural gas reported by four weather reporting stations; W_{H1}: expected temperature shock for heating oil reported by one weather reporting station; W_{H2}: expected weighted average temperature shock for heating oil reported by four weather reporting stations; EINVN: expected inventory shock for natural gas; EINVH: expected inventory shock for heating oil; DJ: Dow Jones industrial index.

Table 2: Results for Ljung-Box Q test

Commodity	Variable	Q(1)	Q(2)	Q(3)	Q(4)	Q(5)	Q(6)	Q(9)	Q(12)
RET 1	Series 1	0.1670	0.3630	1.8291	2.2173	4.1915	4.6245	6.7521	8.0710
		(0.6830)	(0.8340)	(0.6090)	(0.6960)	(0.5220)	(0.5930)	(0.6630)	(0.7800)
	Series 2	0.2228	0.3636	2.0589	2.3091	4.5976	5.1488	7.2407	8.6039
		(0.6370)	(0.8340)	(0.5600)	(0.6790)	(0.4670)	(0.5250)	(0.6120)	(0.7360)
RET 2	Series 1	2.9185	3.9548	4.2424	4.2924	5.4078	5.6102	7.7109	11.3040
		(0.0880*)	(0.1380)	(0.2360)	(0.3680)	(0.3680)	(0.4680)	(0.5640)	(0.5030)
	Series 2	3.1149	4.0195	4.3823	4.4941	5.8186	6.0655	8.1283	11.5900
		(0.0780*)	(0.1340)	(0.2230)	(0.3430)	(0.3240)	(0.4160)	(0.5210)	(0.4790)
RET 1	Series 1	3.9518	4.5093	4.5625	6.0989	6.1009	6.6292	8.5532	10.4580
		(0.0470**)	(0.1050)	(0.2070)	(0.1920)	(0.2970)	(0.3570)	(0.4790)	(0.5760)
	Series 2	3.9350	4.4883	4.5397	6.0974	6.0986	6.6583	8.5968	10.4910
		(0.0470**)	(0.1060)	(0.2090)	(0.1920)	(0.2970)	(0.3540)	(0.4750)	(0.5730)
RET 2	Series 1	3.8889	4.1075	4.2889	5.7237	6.0139	6.7367	8.1229	9.7870
		(0.0490**)	(0.1280)	(0.2320)	(0.2210)	(0.3050)	(0.3460)	(0.5220)	(0.6350)
	Series 2	3.8486	4.0742	4.2522	5.7043	6.0079	6.7678	8.1760	9.8194
		(0.0500)	(0.1300)	(0.2350)	(0.2220)	(0.3050)	(0.3430)	(0.5170)	(0.6320)

Table 3 shows that both Ljung-Box Q^2 and ARCH-LM statistics are significant, suggesting heteroskedasticity. This study, thus, uses the GARCH model rather than the OLS model to estimate and analyze the influence of temperature change shock and inventory surprises on natural gas and heating oil futures returns. Table 4 lists the optimum GARCH results for the influence of the expected temperature shock based on the temperature reported by a single weather reporting station or the weighted average temperature reported by multiple weather reporting stations on energy futures returns. These findings indicate that the log likelihood values from the weighted average temperature reported by multiple weather reporting stations are lower than those from a single weather reporting station for energy futures returns, suggesting that the average weighted temperature data obtained from multiple weather reporting stations are more appropriate for examining the impact of temperature change on energy futures returns than those obtained from a single weather station. That means the average weighted temperature should be used to estimate the GARCH model for energy futures returns. Empirical results show that W_{N1}, W_{N2}, W_{H1}, and W_{H2} are all significantly positive for both near-month and far-month natural gas and heating oil futures returns, suggesting that when the market expects the degree day in the future will be higher (lower) than the average degree day level, the energy demand will increase (decrease), causing near-month and far-month natural gas or heating oil futures returns to increase (decrease). Table 4 also shows that expected inventory surprises of heating oil do not significantly impact heating oil futures returns. However, this study finds that the expected inventory surprises of natural gas are significantly and negatively related with natural gas futures returns, which is consistent with the findings of Ates and Wang [4]. Table 4 also shows that DJ is significantly and negatively associated with heating oil futures returns suggesting that the increase (decrease) of DJ decreases (increases) near-month and far-month heating oil futures returns.

Table 3: Results for Ljung-Box Q^2 and ARCH-LM tests

Commodity	Variables	$Q2$ (3)	$Q2$ (6)	$Q2$ (9)	$Q2$ (12)	ARCH(6)	ARCH((12))
RETN1	Series 1	37.4260	47.8960	48.1930	57.5900	7.4251	4.3999
		(0.0000***)	(0.0000***)	(0.0000***)	(0.0000***)	(0.0000***)	(0.0000***)
	Series 2	37.7920	47.9380	48.2280	57.8320	7.4365	4.4098
		(0.0000***)	(0.0000***)	(0.0000***)	(0.0000***)	(0.0000***)	(0.0000***)
RETN2	Series 1	49.2860	49.7660	50.2800	58.3290	7.0381	4.4487
		(0.0000***)	(0.0000***)	(0.0000***)	(0.0000***)	(0.0000***)	(0.0000***)
	Series 2	48.7730	49.2380	49.7220	57.6940	6.9851	4.4071
		(0.0000***)	(0.0000***)	(0.0000***)	(0.0000***)	(0.0000***)	(0.0000***)
RETH1	Series 1	9.4883	22.1670	29.4190	33.8710	3.1697	2.1361
		(0.0230**)	(0.0010***)	(0.0010***)	(0.0010***)	(0.0043***)	(0.0129**)
	Series 2	9.5893	22.1060	29.3000	33.7370	3.1614	2.1273
		(0.0220**)	(0.0010***)	(0.0010***)	(0.0010***)	(0.0044***)	(0.0133**)
RETH2	Series 1	1.2064	13.6810	22.5980	28.2530	2.3223	2.1579
		(0.7510)	(0.0330**)	(0.0070***)	(0.0050***)	(0.0311**)	(0.0118**)
	Series 2	1.2168	13.4810	22.4570	28.1050	2.2882	2.1480
		(0.7490)	(0.0360**)	(0.0080***)	(0.0050***)	(0.0336**)	(0.0123**)

Table 4: Summary of the near-month and far-month energy futures prices

RET$_{Z1}$ − W$_{Z1}$ Mean Eqn Variables	NN	A single weather station		
		FN	NH	FH
		Coefficients (P value)		
C	0.1464	0.1595	0.0921	0.0933
(P value)	(−0.2258)	(0.1416)	(0.2187)	(0.1938)
W$_{Z1}$	0.062***	0.0488***	0.0315**	0.0093**
(P value)	(−0.0037)	(0.0027)	(0.0389)	(0.0442)
EINVZ	−0.0095*	−0.0141**	−0.0001	0.0000
(P value)	−0.0656	(0.0435)	(0.1354)	(0.2134)
DJ	−0.2378*	−0.1876	−0.3564***	−0.4307***
(P value)	(−0.0985)	(0.2714)	(0.0007)	(0.0000)
Variance Eqn		Coefficients (P value)		
C	0.2375	0.541***	−0.0395	0.0349
(P value)	(−0.5209)	(0.0048)	(0.8130)	(0.8535)
ARCH(1)	0.1273***	0.0230***	0.0460***	0.0246**
(P value)	(0.0000)	(0.0000)	(0.0020)	(0.0244)
GARCH(1)	0.7956***	0.9669***	0.9026***	0.9250***
(P value)	(0.0000)	(0.0000)	(0.0000)	(0.0000)
WIN	0.1474	−0.1791	0.1671*	0.1348*
(P value)	(−0.4743)	(0.4894)	(0.0535)	(0.0723)
IDZ	2.6683*	−2.1726***	1.3818*	0.8908
(P value)	(−0.0329)	(0.0011)	(0.0577)	(0.1852)
STM	1.0145	0.7189	−0.0752	−0.016
(P value)	(−0.1658)	(0.1017)	(0.6786)	(0.4876)
W$_{Z1}$	0.1108***	0.0008**	0.0246**	0.0202**
(P value)	(0.0000)	(0.0164)	(0.0053)	(0.0145)
W$_{Z1}^2$	0.002***	0.0013***	−0.0008	−0.0041
(P value)	(−0.0011)	(0.0021)	(0.5871)	(0.3258)
Sum of coeff	0.9229	0.9899	0.9486	0.9496
Log likelihood	−2661.77	−2591.84	−2273.66	−2207.37

RET$_{Z3}$ − W$_{Z3}$ Mean Eqn Variables	NN	Multiple weather stations		
		FN	NH	FH
		Coefficients (P value)		
C	0.1060	0.1382	0.8816	0.0914
(P value)	(0.3851)	(0.2324)	(0.2406)	(0.2046)
W$_{Z3}$	0.1038***	0.0983***	0.0310*	0.0287*
(P value)	(0.0017)	(0.0020)	(0.0583)	(0.0682)
EINVZ	−0.0095*	−0.0104**	−0.0001	0.0000
(P value)	(0.0738)	(0.0397)	(0.1551)	(0.2390)
DJ	−0.2436	−0.1586	−0.3627***	−0.4365***
(P value)	(0.1020)	(0.2667)	(0.0004)	(0.0000)
Variance Eqn		Coefficients (P value)		
C	0.1262	2.8616***	−0.0822	0.0000
(P value)	(0.7076)	(0.0025)	(0.6111)	(0.9994)
ARCH(1)	0.1071***	0.1288***	0.0431***	0.0247**
(P value)	(0.0000)	(0.0000)	(0.0028)	(0.0196)
GARCH(1)	0.8321***	0.5425***	0.9092***	0.9277***
(P value)	(0.0000)	(0.0000)	(0.0000)	(0.0000)
WIN	0.1495	−0.2403	0.1415	0.1165
(P value)	(0.3728)	(0.4829)	(0.1033)	(0.1029)
IDZ	2.1381*	2.7170***	1.4443**	0.9592
(P value)	(0.0839)	(0.0006)	(0.0446)	(0.1491)
STM	1.0828	1.7002	−0.0616	−0.0755
(P value)	(0.1036)	(0.1036)	(0.7282)	(0.5659)
W$_{Z3}$	0.1186***	0.1296**	0.0278**	0.0227**
(P value)	(0.0000)	(0.0272)	(0.0026)	(0.0103)
W$_{Z3}^2$	0.0169***	0.0381***	−0.0003	−0.0007
(P value)	(0.0031)	(0.0017)	(0.8652)	(0.6057)
Sum of coeff	0.9391	0.6713	0.9524	0.9524
Log likelihood	−2662.70	−2604.80	−2273.78	−2207.65

***1% significance level; **5% significance level; *10% significance level. NN: near-month natural gas; NH: near-month heating oil; FN: far-month natural gas; FH: far-month heating oil.

Temperature in Table 4 is reported by single and multiple weather reporting stations. Regarding the variance equations, this study finds data for the winter heating season, as listed in Table 4, is mostly insignificant, suggesting that energy futures returns are generally not higher during the winter than other seasons, which is inconsistent with the findings in [5]. The adjusted R-squares range for the temperature reported by one weather reporting station is from 0.003 to 0.027; that range for the temperature reported by multiple weather reporting stations is 0.005 to 0.026. Empirical results show that both natural gas and heating oil's announcement day is significantly positive, suggesting that a storage report announcement effect exists, which is consistent with the literature. This work finds that hurricane announcement effect does not exist for energy futures returns.

This investigation also finds significantly positive nonlinear temperature effects for natural gas futures returns, suggesting the market expects degree day volatility to increase with natural gas demand. Furthermore, since the sum of GARCH coefficients can measure the persistence of market volatility, this study finds that both the sum of the coefficients affected by the temperature reported from a single weather reporting station and that from weighted average temperature of multiple weather reporting stations approaches one, except for the far-month natural gas futures prices, consistent with the stable convergence of variance in the GARCH model.

CONCLUSIONS

We used the GARCH model to examine the influence of the unexpected temperature shock based on both the temperature reported by a single weather reporting station and the weighted average temperature reported by multiple weather reporting stations, expected inventory surprise, the movement of Dow Jones Industrial Index (DJ), winter heating season, storage announcements, hurricane announcements, and nonlinear effect of temperature on near-month and far-month energy natural gas and heating oil futures returns. Several interesting results are summarized as follows.

- The expected temperature shock significantly and positively affects near-month and far-month natural gas and heating oil futures returns. The implication is that, along with the drastic change in temperatures over past four decades, the uncertainly of economic environmental factors increases, increasing the volatility of natural gas and heating oil prices and raising the required rate of returns on natural and heating oil futures. This suggests that the investors should invest a portion of their funds on the weather derivative commodities for their hedging and arbitraging purposes.

- Significant temperature effect on both the conditional mean and volatility of both natural gas and heating oil futures prices, suggesting that when the market expects the degree days in the future to be higher than average, energy demand increase, increasing the near-month and far-month natural gas and heating oil futures returns. This result confirms with the finding of Mu [5] for natural gas futures.

- Expected inventory surprises significantly and negatively impacts the far-month natural gas futures returns, confirming the finding in [10, 21]. However, the expected inventory surprises do not affect heating oil futures returns.

- Expected nonlinear temperature effect exists for natural gas futures returns, suggesting the market expects the degree day volatility to increase with natural gas demand. However, nonlinear temperature effects does not exist for heating oil futures returns.

- DJ returns significantly and negatively affect heating oil futures returns.

- Storage announcement for natural gas significantly affects near-month and far-month natural gas futures returns; while storage announcements for heating oil significantly affect near-month heating oil futures returns. These results show that information about market fundamentals is an important determination of near-month energy futures returns.

- The hurricane announcement and winter heating season do not significantly affect natural gas and heating oil futures returns; the latter is inconsistent with the findings in [4, 5].

- Using the GARCH, the results indicate that the weighted average temperature data obtained from multiple weather reporting stations is more appropriate than that got from a single weather reporting station for examining the impact of the temperature shock on natural gas and heating oil futures returns.

LIMITATIONS AND FUTURE RESEARCH DIRECTIONS

This study has the following restrictions.

- The temperature data of US weather stations at each state is difficult for collection, causing the sample period of this study to be limited to four years.

- Since a large portion of the near-month and the far-month's energy products data are hard to collect, the research targets of this study regarding energy commodities are limited to natural gas and heating oil only.

- Since most of the previous researches examined the impact of the temperature on the agricultural commodity futures, only a few studies examined that on the energy product futures; the relevant references for this investigation are limited.

This investigation suggests future researchers to examine the impact of temperature on other energy commodities in addition to the natural gas and heating oil or to include longer sample period than four years for various energy products futures returns.

ACKNOWLEDGMENTS

The authors would like to thank the precious remarks provided by Xiaoyi Mu and the participants in the 87th WEA conference (Anita Chaudhry, Robert Tischer, Weigui Yu, and Jy Si Wu). Ted Knoy is also appreciated for his editorial assistance. Sincere thanks are given to the anonymous referees for their valuable comments. This research is partially supported by the National Science Council of Taiwan under Grant NSC 102-2410-H-033-039-MY2.

REFERENCES

1.   M. Ruth, D. Coelho, and D. Karetnikov, "The U.S. economic impacts of climate change and the costs of inaction," in A Review and Assessment by the Center for Integrative Environmental Research (CIER), University of Maryland, 2007.

2. G. Considine, Introduction to Weather Derivative, Weather Derivatives Group, Aquila Energy, 2000.

3. S. C. Stevens, "Evidence for a weather persistence effect on the corn, wheat, and soybean growing season price dynamics," Journal of Futures Markets, vol. 11, no. 1, pp. 81–88, 1991.

4. A. Ates and G. H. K. Wang, "Price dynamics in energy spot and futures markets: the role of inventory and weather," in Proceedings of Financial Management Association Annual Meeting, Vienna, Austria, 2007.

5. X. Mu, "Weather, storage, and natural gas price dynamics: Fundamentals and volatility," Energy Economics, vol. 29, no. 1, pp. 46–63, 2007.

6. K. C. Chen, R. S. Sears, and D. N. Tzang, "Oil prices and energy futures," Journal of Futures Markets, vol. 7, no. 5, pp. 501–518, 1987.

7. J. H. Herbert, "Trading volume, maturity and natural gas futures price volatility," Energy Economics, vol. 17, no. 4, pp. 293–299, 1995.

8. W. D. Walls, "Econometric analysis of the market for natural gas futures," The Energy Journal, vol. 16, no. 1, pp. 71–83, 1995.

9. M. D. Chinn, M. LeBlanc, and O. Coibion, "The predictive content of energy futures: an update on petroleum, natural gas, heating oil and gasoline," NBER Working Paper 11033, 2005.

10. S. Z. Chiou-Wei, C. L. Scott, Z. Zhu, and C. H. Guernsey, "The response of U.S. natural gas futures and spot prices to storage change surprises and the effect of escalating physical gas production," Working Paper, 2013.

11. H. Suenaga, A. Smith, and J. Williams, "Volatility dynamics of NYMEX natural gas futures prices,"Journal of Futures Markets, vol. 28, no. 5, pp. 438–463, 2008.

12. M. M. Ghiasi, A. Bahadori, and S. Zendehboudi, "Estimation of triethylene glycol (TEG) purity in natural gas dehydration units using fuzzy neural network," Journal of Natural Gas Science and Engineering, vol. 17, pp. 26–32, 2014.

13. A. Aramesh, N. Montazerin, and A. Ahmadi, "A general neural and fuzzy-neural algorithm for natural gas flow prediction in city gate stations," Energy and Buildings, vol. 72, pp. 73–79, 2014.

14. M. Kovacic and B. Sarler, "Genetic programming prediction of the natural gas consumption in a steel plant," Energy, vol. 66, pp. 273–284, 2014.

15. G. M. Ljung and G. E. P. Box, "On a measure of lack of fit in time series models," Biometrika, vol. 65, no. 2, pp. 297–303, 1978.

16. T. Bollerslev, "Generalized autoregressive conditional heteroskedasticity," Journal of Econometrics, vol. 31, no. 3, pp. 307–327, 1986.

17. D. A. Dickey and W. A. Fuller, "Distribution of the estimators for autoregressive time series with a unit root," Journal of the American Statistical Association, vol. 74, no. 366, pp. 427–431, 1979.

18. D. A. Dickey and W. A. Fuller, "Likelihood ratio statistics for autoregressive time series with a unit root," Econometrica. Journal of the Econometric Society, vol. 49, no. 4, pp. 1057–1072, 1981.

19. D. Gujarati, Basic Econometrics, McGraw-Hill, Singapore, 1995.

20. R. F. Engle, "Autoregressive conditional heteroscedasticity with estimates of the variance of United Kingdom inflation," Econometrica, vol. 50, no. 4, pp. 987–1007, 1982.

21. H. Geman and S. Ohana, "Forward curves, scarcity and price volatility in oil and natural gas markets,"Energy Economics, vol. 31, no. 4, pp. 576–585, 2009.

Chapter 2

Regulatory Nirvana for Hydraulic Fracture Stimulation

Barry Goldstein[1], Michael Malavazos[1], Alexandra Wickham[1], Michael Jarosz[1], Dominic Pepicelli[1], Mieka Webb[1], and Dale Wenham[1]

[1]Energy Resource Division, 'Department for Manufacturing, Innovation, Trade, Resources & Energy (DMITRE), State Government of South Australia, Australia

ABSTRACT

Government are challenged to deploy trustworthy regulation to enable profitable and environmentally sustainable unconventional petroleum projects. A key activity under scrutiny during the development of these projects is hydraulic fracture stimulation. Regulatory 'Nirvana' for unconventional projects and conventional projects alike entails:

- Pragmatic licence tenure;
- Regulatory certainty and efficiency without taint of capture;
- Regulators and licensees with trustworthy competence and capacity;
- Effective stakeholder consultation well-ahead of land access;
- Public access to details of significant risks and reliable research to backup risk management strategies so the basis for regulation is contestable anytime, everywhere;
- Timely notice of entry with sufficient operational details to effectively inform stakeholders;
- Potentially affected people and organisations can object to land access - without support for vexatious objections;
- Fair and expeditious dispute resolution processes;
- Fair compensation to affected land-users;
- Risks are reduced to low or as low as reasonably practicable (ALARP) while also meeting community expectations for net outcomes;
- Licensees monitor and report on the efficacy of their risk management, and the regulator probes same;
- Regulator can prevent and stop operations, require restitution, levy fines and cancel licences; and
- Industry compliance records are public, so the efficacy of regulation is transparent.

These principles are deployed in South Australia where:

- 24 unconventional gas plays are being explored, each with giant gas potential;
- Hundreds of wells have been safely hydraulically fracture stimulated;

Since implementing South Australia's Petroleum and Geothermal Energy Act 2000 [1] (PGE Act), more than 11,000 notices of entry for petroleum operations led to just one court action, and that was to establish a legal precedent that geophysical surveys can extend outside a licence to enable a complete understanding of the potential resources within a licence.

The introduction of new energy development technologies is inevitable, so regulatory Nirvana requires adaptive learning so that the previously mentioned principles are maintained. Expeditious, welcomed access to land for compatible, multiple uses is the metric for performance, and leading practice is based on the principle that trust is the most valuable lead factor and lag outcome in sustaining land access for resource exploration, development and production.

INTRODUCTION

The Australian oil and gas industry has contributed greatly to the economic prosperity and quality of life of our communities for decades to date. An opportunity to prolong and expand welcomed contributions in a golden age of unconventional gas is arising. The challenges ahead of a prospective golden age of unconventional gas are many, and include getting regulation and operations right. Results that consistently, simultaneously meet community and investor expectations for social, environmental and economic outcomes will deliver trust in land access and investment – and create a virtuous lifecycle for the upstream petroleum sector for decades to come.

Coal seam methane was Australia's first unconventional gas play to be commercialised and reserves will underpin LNG exports from Gladstone, Queensland. In October 2012, the tap was turned on the first domestic commercial use of shale gas from Moomba 191 in the Cooper Basin – another milestone on the road to develop a variety of unconventional gas resources across Australia. Foreseeing the potential scope of development of unconventional gas resources:

- Companies have shifted budgets to explore, appraise and develop unconventional gas plays;
- People and organisations potentially affected by unconventional gas operations have justifiably expressed concerns for preserving social, natural and economic environments; and
- Governments have made strides to refine regulatory and investment settings to simultaneously satisfy both community and investor expectations for net outcomes.

In this regard, October 2010, the South Australian Government's Department for Manufacturing, Innovation, Trade, Resources and Energy (DMITRE) initiated a consultative group to inform how unconventional gas projects could be undertaken the most sustainably and efficiently, considering the social, environmental and economic impacts and benefits. This group – the Roundtable for Unconventional Gas Projects in South Australia (Roundtable) – played a critical role, informing our Roadmap for Unconventional Gas Projects in South Australia (Roadmap) [2]. As of January 2013, the Roundtable had 230 members including peak representative bodies, companies, universities, media outlets, individuals and key government agencies from all the states, the Northern Territory and the Commonwealth governments. This paper summarises the findings of this Roadmap that relate to world leading practices for the regulation of the development of unconventional petroleum resources that rely on hydraulic fracture stimulation to attain economic flow rates.

THE ROADMAP

The Roadmap for Unconventional Gas Projects in South Australia [2] was developed to provide timely, credible information to people, communities and markets, outlining potential risks and rewards associated with unconventional gas projects. It sets the course for the environmentally sustainable development of South Australia's large endowment of unconventional gas, and encourages safe exploration and production under this State's robust and effective regulatory framework, the PGE Act. The Roadmap helps to ensure people and enterprises potentially affected by unconventional gas projects understand the regulatory framework, the transparent environmental assessment and activity approval processes; and how they will be consulted, so their rights to object in part or in full are supported. The Roundtable also identifies 125 recommendations which cover the life cycle of unconventional gas projects – from exploration to production and possible liquefied natural gas exports, as well as related supply chains and infrastructure matters. Roundtable working groups have reconvened to develop plans to implement these recommendations.

To comment on and further inform the implementation of the 125 recommendations posed in the Roadmap or to enquire regarding participation in the Roundtable for Unconventional Gas Projects in South Australia – readers are asked to contact dmitre.petroleum@ sa.gov.au.

REGULATION TO ENABLE HYDRAULIC FRACTURE STIMULATION IN THE PUBLIC'S INTEREST — THE SOUTH AUSTRALIAN APPROACH

Onshore petroleum exploration and development activities in South Australia are administered by DMITRE under the South Australian PGE Act. The PGE Act has a number of aspects that are considered a comparative advantage without precedent in other Australian legislation [3]. High level objectives of the PGE Act include:

- Sustain trusted practical, efficient, effective and flexible regulation for upstream petroleum, geothermal and gas storage enterprises, and the construction and operation of transmission pipelines, in the State;
- Encourage and maintain competition in the upstream petroleum and geothermal sectors;
- Minimise environmental damage and protect the public from risks inherent in petroleum and geothermal operations;
- Sustain effective consultation processes with people affected by regulated activities, and the public in general; and
- Ensure as far as reasonably practical the security of supply of natural gas.

It is important in this discussion to highlight that in the context of the PGE Act the definition of environment (under s. 4 of the PGE Act) is broad, and includes:

- Land, air, water (including both surface and underground water)
- Organisms and ecosystems – this includes native vegetation and fauna;
- Buildings, structures and cultural artefacts;
- Productive capacity or potential;
- The external manifestations of social and economic life which includes aspects such as human health and wellbeing; and
- The amenity values of an area.

This definition of environment is consistent with the Environment Protection Act 1993 [4] definition, and is broad to ensure that potential impacts on all natural, social and economic aspects of the environment are identified, considered, and appropriately addressed through the environmental assessment and approval provisions of the PGE Act.

A key lesson learnt by DMITRE in post-event investigations of significant incidents is that regulators must have relevant and up-to-date capabilities (competence and capacity) to be trusted to act in the interests of the many stakeholders involved in upstream petroleum industry activities. This includes protecting natural, social and economic environments; effectively managing the risks of regulatory capture [5]; and providing expeditious approvals. As the regulator of upstream petroleum and geothermal energy activities in South Australia, administering the PGE Act, DMITRE strives to maintain a one-stop-shop or lead agency approach.

This approach has been discussed by Australia's Productivity Commission [6] which concluded:

- One-stop-shops (lead agencies) are the most efficient regulatory approach when well managed without capture;
- Under a lead agency approach … approval of most, if not all, aspects of an application would rest with one designated agency. This agency …would maintain control of the process and in most cases, would consult with other relevant agencies, such as an environmental agency, rather than formally refer the application to a separate agency for assessment. In some limited circumstances where impacts are considered to be significant, a formal referral may take place. By maintaining control of the approval process the lead agency approach is able to streamline approval processes and minimise time delays.
- South Australia's one-stop-shop (through DMITRE), 'is widely seen as a model for other jurisdictions to emulate';
- With appropriate governance, experience in South Australia suggests that [lead agencies] can achieve an appropriate balance between enforcing legislative provisions and expediting approvals.

Properly resourced one-stop-shops (lead agencies) transparently facilitate the delivery of all co-regulatory objectives and requirements, and hence earn trust from the industry, co-regulatory agencies and the public. A one-stop-shop approach enables stewardship of approval processes in parallel rather than in series.

Through this approach DMITRE works closely with its co-regulatory agencies, such as the South Australian Environment Protection Authority (EPA), Department of Environment, Water and Natural Resources (DEWNR), SafeWork SA, Department of Health, Department of Planning, Transport and Infrastructure (DPTI) and Aboriginal Heritage to deliver an efficient application of all relevant laws and regulations applicable to the petroleum and geothermal industries in South Australia.

The PGE Act has been designed to enable a one-stop shop approach such that in complying with the objectives of the PGE Act, upstream petroleum operations' compliance with obligations under other legislation will also be facilitated. These concurrent legislation and requirements include:

- The Commonwealth's Environmental Protection, Biodiversity and Conservation Act 1999 (EPBC Act) internationally important flora, fauna, ecological communities and heritage places — defined in the EPBC Act as matters of national environmental significance. The Commonwealth Government Department of Sustainability, Environment, Water, Population and Communities (SEWPaC) provides stewardship for the EPBC Act;

- South Australia's Environment Protection Act 1993 (EP Act), and relevant policies that provide the regulatory framework to protect South Australia's environment, including land, air and water. This legislation was the result of the streamlined integration of six Acts of Parliament and the abolition of the associated statutory authorities. South Australia's EPA provides stewardship for this Act;

- South Australia's National Parks and Wildlife Act 1972 (NP&W Act), which is the cornerstone for protecting natural environments within parks and regional reserves in the State. The DEWNR provides stewardship for this Act. The NP&W Act is significant as it is a key part of the co-regulatory approval regime for minerals and energy (including unconventional gas) resource exploration and production in South Australia;

- The South Australian Work Health and Safety Act 2012 (SA) (WHS Act) is the state's lead legislation to protect people in the workplace. SafeWorkSA provides stewardship for this Act;

- The South Australian Native Vegetation Act 1991 (NV Act), administered by DEWNR;
- The South Australian Natural Resources Management Act 2004 (NRM Act), administered by DEWNR;
- The South Australian Development Act 1993, administered by the DPTI;
- The South Australian Public and Environmental Health Act 1987, and specifically the Public and Environmental Health (Waste Control) Regulations 2010, as administered by HealthSA
- The Native Title (South Australia) Act 1994, administered by the State's Attorney General's Department
- The Commonwealth Native Title Act 1993 (NT Act) administered by the Commonwealth's Attorney General's Department
- The South Australian Adelaide Dolphin Sanctuary Act 2005, administered by DEWNR
- The South Australian Aboriginal Heritage Act 1988 administered by the State's Department of Aboriginal Affairs and Reconciliation
- The South Australian Marine Parks Act 2007 administered by DEWNR
- The South Australian River Murray Act 2003 administered by DEWNR; and
- The South Australian Arkaroola Protection Act 2012 administered by DEWNR.

Compliance with these pieces of legislation is facilitated through collaborations and working arrangements between DMITRE and the government agencies that administer these Acts, to ensure that the Statements of Environmental Objectives (SEO) that must be complied with for specific activities are consistent and in keeping with the relevant objects of each of these Acts.

PRINCIPLES FOR BEST PRACTICE REGULATION

The PGE Act was developed on the basis of the following 6 principles for regulatory best practice:

- *Certainty:* The regulatory objectives are uniform, clear, and predictable for all stakeholders.
- *Openness:* Stakeholders are appropriately consulted on the establishment of the regulatory objectives.
- *Transparency*: The regulatory decision-making processes are visible and comprehensible to all stakeholders and industry performance in terms of compliance with the regulatory objectives is clear to all stakeholders.
- *Flexibility*: The level of regulatory scrutiny, surveillance and enforcement needed to ensure compliance is determined on the basis of individual company compliance capability and the outcomes to be achieved.
- *Practicality*: The regulatory objectives are achievable and measurable.
- *Efficiency*: The compliance costs imposed on both government and the licensee by the regulatory requirements are minimised and justified. Negative impacts on communities are minimised, and licensees remain liable for the cost of their impacts. Furthermore, an appropriate rent (Royalty) is paid to the community from the value realised from the development and production of its natural resources.

The above listed Regulatory Principles can be achieved through the following regulatory strategies.

- Regulatory objectives and assessment criteria for those objectives are developed through broad stakeholder consultation involving industry, government agencies and the community to ensure acceptance and credibility in the environmental objectives to be achieved
- Regulators and licensees maintain trustworthy capabilities (competence and capacity)
- Effective, informative stakeholder consultation by both project operators and regulators is initiated well ahead of land access. This drives operators to explain their planned activities and any potential risks, seek feedback on areas of interest or concern for the community, and establish relationships and terms for land access with stakeholders well before applying for activity approval from DMITRE, e.g. before any particular activity 'gets personal'

- Provide public access to details of risks, reliable research to reduce key uncertainties and support risk management strategies so the basis for regulation is contestable
- Timely notice of entry with sufficient operational details to effectively inform stakeholders
- Potentially affected people and organisations can object to land access – while the regulator and prescribed dispute resolution processes do not support, and hence minimise, vexatious objections
- Fair and expeditious dispute resolution processes
- Fair compensation to affected land-users for costs, losses, and deprivation of land use due to operations
- Reduction of risks to low or as low as reasonably practicable (ALARP), while also meeting community expectations for overall outcomes
- Licensees monitor and report (to the regulator) on the efficacy of their risk management processes, and the regulator probes same
- The regulator can prevent and stop operations, require restitution or rehabilitation, levy fines and cancel licences
- Industry compliance records are made public, so the efficacy of regulation is transparent.

Clear, efficient and effective activity approval processes are fundamental for trustworthy regulation. Mapping approval processes can also elucidate scope for increased efficiency and reduced red tape. Figures 1, 2 and 3 illustrate the three-stage process for petroleum and geothermal licensing and approvals in South Australia with a one-stop-shop approach led by DMITRE, for exploration, retention, production and associated activities.

The first stage (Figure 1) entails the grant of a licence authorising the licensee to carry out specific activities to which the licence relates. Environmental assessments are required in the second stage (Figure 2). Statements of Environmental Objectives (SEOs) and environmental assessment criteria for activity approvals are established in this second stage. Finally, in the third stage (Figure 3), a location-specific activity notification is submitted for assessment and approval, where required.

All three stages are required to be completed before regulated activities can commence. In practice, it is possible for some aspects of each stage to progress in parallel. This flexibility is most easily enabled through discussions with the regulator (DMITRE) early in the planning process. Figures 1, 2 and 3 specify relevant regulations (of the PGE Act) to help guide licensees through these stages.

Figure 1: of licensing and approval process for exploration, retention and production activities pursuant to South Australia's Petroleum and Geothermal Energy Act 2000. (Blue box = initiated by proponent/Licensee and Green box = initiated by DMITRE/ SA Government).

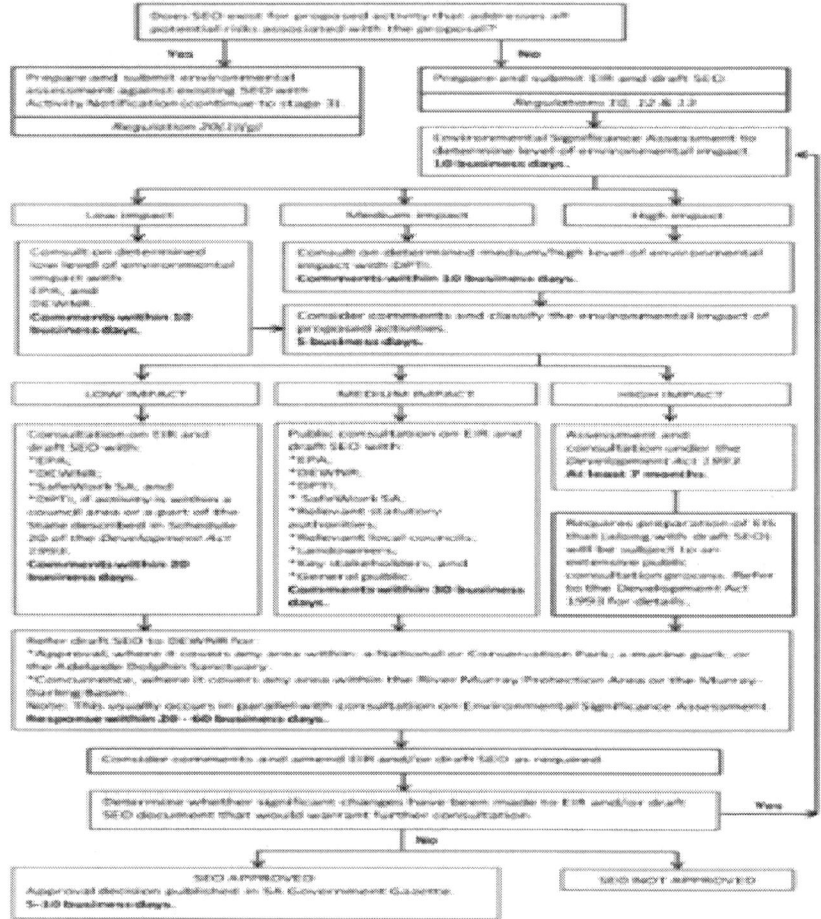

Figure 2: Stage 2 of licensing and approval process for exploration, retention and production activities pursuant to South Australia's Petroleum and Geothermal Energy Act 2000. (Blue box = initiated by proponent/Licensee and Green box = initiated by DMITRE/ SA Government).

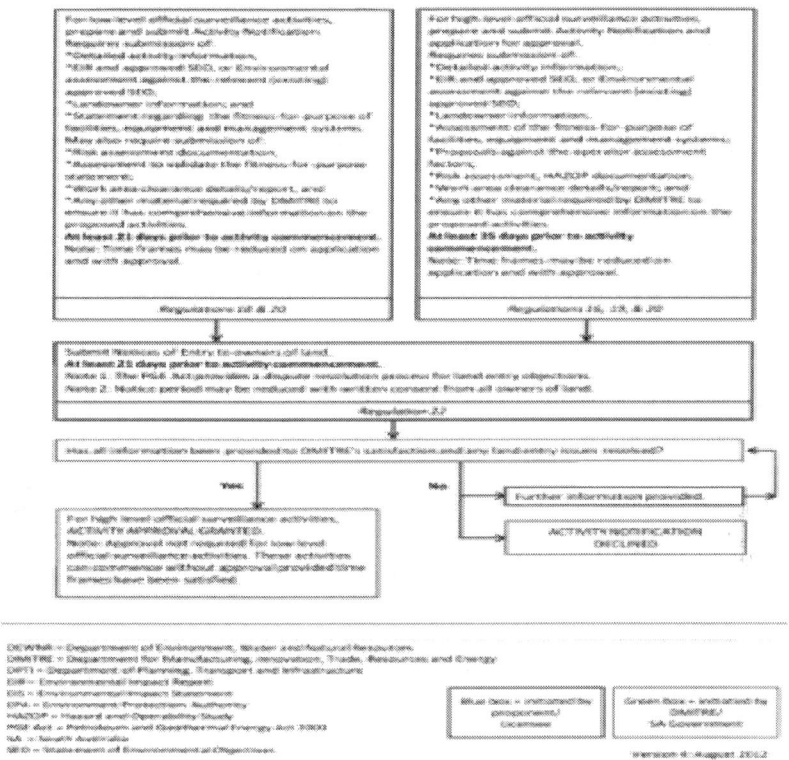

Figure 3: Stage 3 of licensing and approval process for exploration retention and production activities pursuant to the Petroleum and Geothermal Energy Act 2000.

The proceeding sections describe stage 2 (Figure 2) and stage 3 (Figure 3). For details of license authorisation (stage 1) – refer to [2].

ENVIRONMENTAL ASSESSMENT AND APPROVAL

The grant of a PGE Act licence does not provide an automatic entitlement to conduct operations. Rather, regulated activities under the PGE Act (under s. 96) may not be carried out unless an approved Statement of Environmental Objectives (SEO) is in place, prepared on the basis of an Environmental Impact Report (EIR).

The EIR identifies all potential impacts and their risks relating to the activity and the proposed risk mitigation strategies. The SEO identifies the environmental objectives to be achieved to address the risks identified in the EIR and the criteria to be used to assess achievement of the objectives.

Through the consultation requirements of the PGE Act, DMITRE expects that licensees will initiate consultation with stakeholders, generally through information sessions or meetings prior to and during the development of their EIR and SEO, to describe their planned activities and the potential impacts, positive or otherwise, which may be experienced by the stakeholders. This is also an opportunity for the licensee to respond to any queries that their stakeholders may have and to understand stakeholder concerns, to ensure that they are addressed within the EIR and SEO.

Other agencies with the duty of care for ensuring the objects of the legislation that they administer are met are also consulted early to ensure their requirements are included within the objectives detailed in the SEO.

Once an EIR and draft SEO have been prepared and submitted for assessment, DMITRE uses the information provided in the EIR to complete an environmental significance assessment to determine the level of environmental impact of the activity. If prior consultation is not demonstrated, then DMITRE will conduct a broader consultation on the draft documents to ensure stakeholders including landholders and other government departments have been provided with opportunities to raise any issues of concern they may have with the activities as described or the level or accuracy of information provided, prior to SEO approval and well before the commencement of regulated activities.

The significance assessment is conducted based on the information provided in the EIR and in accordance with publicly documented criteria to assess the level of certainty in the predicted impacts, their potential consequences related to the proposed activities and the degree to which these consequences can be managed. The environmental significance criteria include assessment of the level of stakeholder concern. In cases where the level of stakeholder consultation is not demonstrated or the EIR documents high levels of stakeholder concern then this may indicate deficiencies in stakeholder consultation during the development of the EIR and draft SEO. Where

DMITRE's assessment identifies such a deficiency, the determined level of environmental significance may be greater and likely to trigger more extensive stakeholder consultation by DMITRE. This ensures relevant stakeholders are provided with appropriate time for opinions to be considered and represented equitably in advance of SEO and subsequent activity approvals.

The combination of the outcomes of the significance assessment criteria lead to the determination of an overall level of environmental impact of the activity as low, medium or high. The level of environmental impact that is assigned to a particular activity in turn determines the consultation that DMITRE undertakes, both on the level assigned, and the content of the EIR and draft SEO documents. These consultation arrangements are outlined within the PGE Act, and within administrative arrangements between DMITRE and its co-regulatory agencies, which are all available on the DMITRE website [7].

Regardless of the determined level of environmental impact, all SEOs and associated EIRs are public documents and can be found on the DMITRE website [8] within the Activity Reports section of the Environmental Register.

ACTIVITY NOTIFICATION AND APPLICATION FOR APPROVAL

The grant of PGE Act petroleum exploration, retention, production and pipeline licences does not provide an automatic entitlement to land access for regulated upstream petroleum operations.

Once an SEO is approved, a licensee can apply for approval to undertake a specific activity that is described within the relevant EIR and SEO. With the activity approval application the licensee provides DMITRE with an Activity Notification (Regulation 20 of the PGE Act) which contains detailed activity information including:

- an environmental assessment of the activity against the relevant SEO, including assessment as to whether the activity may have potential significant impacts on Matters of National Environmental Significance (MNES)

- landowner information (including copies of notices of entry sent to landowners)
- an assessment of the fitness for purpose of the licensee management systems and any facilities or equipment to be used
- work area clearance details and report
- risk assessment documentation
- any further information or material as required by DMITRE to ensure that the department has comprehensive information on the proposed activities.

Where MNES are identified, then referral to the Commonwealth Minister for Environment will be made by the licensee or the Department, for assessment and a decision as to whether the activity requires approval under the Environment Protection and Biodiversity Conservation Act 1999 (EPBC Act) [9]

Licensees can be classified as carrying out activities requiring high or low level official surveillance. The level of official surveillance determines the information that must be provided in the notification, the level of scrutiny that DMITRE applies during review of the notification, and the period of notice prior to the proposed commencement of activities. The PGE Act outlines operator assessment factors (Regulation 16 of the PGE Act) that consider the licensees policies, procedures, management systems and track record to classify the licensee's level of official surveillance.

NOTICE OF ENTRY

Mutual trust for compatible, sustainable land access for upstream petroleum operations are traditionally indemnified with formal land access agreements struck between licensees, potentially affected people and enterprises. To provide impetus for fair and sustainable land access for petroleum, geothermal energy and gas storage operation in the State, the PGE Act was amended in 2009 to expand the 'owner of land' definition to cover all persons who may be directly affected by regulated activities, entitling them to notices of entry and compensation. This amendment has proved to be a driver for mutual respect. With this incremental legislated requirement, owners of land are provided with opportunities to raise concerns prior to the commencement of regulated activities.

Landowners are provided with information on the nature of the activities to be carried out including any anticipated events and the management of their consequences to minimise risks to an acceptable level, to enable the landowner to make informed decisions on whether this would have an impact on the land.

Landowners are entitled to object to the licensees proposed entry by giving notice to the licensee within 14 days of the licensee notice of proposed entry and the activity cannot be undertaken until the dispute is resolved. The licensee and the landowner should attempt to reach an agreement of terms under which the licensee may enter the land, or if the risks of the activity to the landowner are too high the licensee may choose to modify the activity and re-issue the Activity Notification. Landowners may also raise any issues or concerns associated with the conduct of activities with DMITRE. In rare cases where the licensee and the landowner cannot resolve the dispute, then the Minister may attempt to mediate between the parties or either party may apply to the Warden's court for resolution. To date, disputed Notices of Entry have been resolved through satisfactory negotiation and have not reached the Warden's Court.

Also, under the PGE Act, owners of land are entitled to appropriate compensation from petroleum licensees for any losses, deprivation or reasonable costs sustained during both the process of negotiating land access and for the full period of land access, right through to the decommissioning of any facilities.

COMPLIANCE AND ENFORCEMENT

DMITRE continuously monitors licensee performance and compliance with the PGE Act. South Australia's approach to provide fair, predictable and trustworthy regulation has been described by Malavazos [10] and entails a publicly available compliance policy [11] which is available on the DMITRE website. South Australia's compliance policy is centred on the prevention of harmful incidents, however depending on the severity of an incident may culminate in prosecution and licence cancelation when warranted. The compliance policy is summarised as a compliance pyramid as shown below in Figure 4.

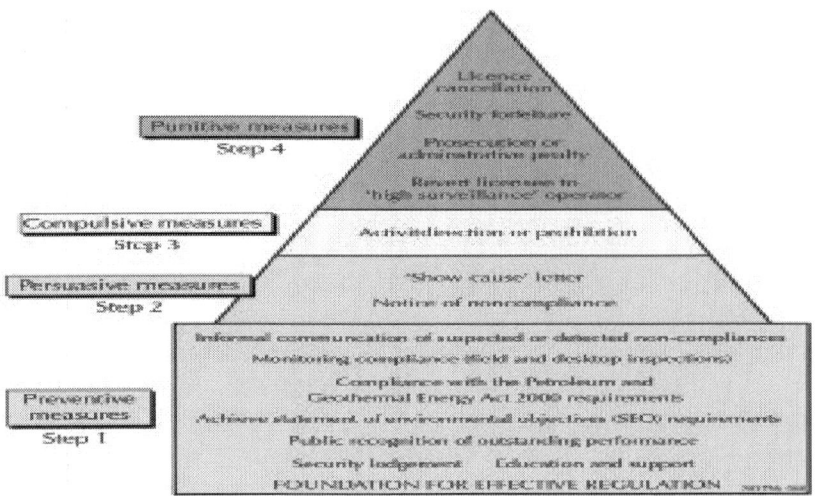

Figure 4: South Australia's compliance enforcement policy under the PGE Act.

DMITRE prepares a PGE Act Annual Compliance Report for the purpose of outlining:

- The compliance monitoring and surveillance activities carried out by DMITRE during each year for activities regulated under the PGE Act;
- Providing an overview of the regulatory performance of the petroleum and geothermal industries in accordance with the requirements of the PGE Act;
- All serious incidents that may have occurred from the previous year; and
- Persuasive, compulsive and punitive enforcement actions that may have been taken during the year (as indicated in Figure 4)

DMITRE's Petroleum and Geothermal Energy Act Compliance Report [12] and Company Annual Reports [13] which report on activities undertaken within each licence area are all publicly available through DMITRE's website.

As well as information provided through the Activity Notifications, DMITRE regularly meets with licensees to discuss their activities and compliance, and conducts ongoing monitoring and surveillance through both field and desktop studies.

CONCLUSIONS

Salient findings from the Roadmap [2] and key aspects of South Australia's current regulation of unconventional gas development, including the regulation of hydraulic fracture stimulation, are summarised below.

- Trusted land access is the most valuable lead factor and outcome.

- Operators and regulators must act early to effectively engage and inform stakeholders so they can make informed decisions on activities. This engagement is best initiated well ahead of land access. South Australia's regulatory framework drives operators to explain their planned activities and any potential risks, seek feedback on areas of interest or concern for the community, and establish relationships and terms for land access with stakeholders well before applying for activity approval from regulators.

- Regulation for compatible, multiple use of land in Australia is undertaken with both risks and net benefits in mind. Considerable net benefits flow from community ownership of subsurface resources when development effectively manages risks to social, natural and economic environments.

- Operators and regulators should adhere to the golden rules for the golden age of gas, as published by the International Energy Agency [14]– which are:
  1. Measure, disclose, engage;
  2. Watch where you drill;
  3. Isolate wells – protect against leaks;
  4. Treat water responsibly;
  5. Eliminate venting and minimise flaring;
  6. Think big; and
  7. Consistent high environmental performance

- International standards [15] for unconventional gas resource and reserve definitions should be adopted.

- Effective, trusted regulation and attractive investment settings are the most effective inputs from governments to beget safe, secure, and competitively priced gas for domestic and international gas markets for decades to come.

- Regulators must have relevant and up-to-date capabilities (competence and capacity) to be trusted to act in the interests of the public in protecting natural, social and economic environments in relation to the full-cycle of mineral and energy resource projects, including unconventional gas operations.
- New energy development technologies will necessitate evolutionary improvement to regulatory frameworks, and best practice regulation will continually evolve
- A one-stop-shop (lead agency) approach to regulation enables co-regulators to do their jobs in parallel, rather than in series. This fosters efficiency without reducing stringent standards for ecologic, social, heritage and economic outcomes.
- Welcomed investment in the development of unconventional gas will effectively reduce risks to as low as reasonably practical while simultaneously meeting community expectations for net outcomes. This will be achieved with, amongst other actions, astute investment in economic unconventional plays,
- The key ingredients of best practice regulation are frameworks that: elicit community trust and investor confidence; provide certainty; entail robust public consultation processes; are transparent; enable flexibility; are open to amendment; are efficient; are practical; and focus on outcomes. This amounts to an overall check-list for best practice co-regulation.

ACKNOWLEDGEMENTS

The authors thank the participants in the Roundtable for Unconventional Gas Projects in South Australia for their valuable advice in developing the Roadmap for Unconventional Gas Projects in South Australia.

REFERENCES

1.   Petroleum and Geothermal Energy Act 2000South Australiahttp://www.legislation.sa.gov.au/LZ/C/A/PETROLEUM%20 AND%20GEOTHERMAL%20ENERGY%20ACT%202000/ CURRENT/2000.60.UN.PDF

2.   Department for Manufacturing, Innovation, Trade, Resources and Energy. Roadmap for Unconventional Gas Projects in South Australia; December 2012http://www.petroleum.dmitre.sa.gov.au/SA_Unconventional_Gas_roadmap

3.   B. A Goldstein, E Alexander, D Cockshell, M Malavazos, J Zabrowarny, The Virtuous Life-Cycle for Exploration and Production (E&P): Lead and Lag Factors. APPEA Journal; 200747

4.   Environmental Protection Act 1993South Australiahttp://www.legislation.sa.gov.au/LZ/C/A/ENVIRONMENT%20PROTECTION%20ACT%201993/CURRENT/1993.76.UN.PDF

5.   M Malavazos, A Model for Environmental and Health and Safety Regulation for the Mining and Upstream Petroleum Industries. Masters thesis. Flinders University South Australia; 1998

6.   Productivity Commission Research Report- Review of Regulatory Burden on the Upstream Petroleum (Oil and Gas) Sector; 2009www.pc.gov.au/__data/assets/pdf_file/0011/87923/upstream-petroleum.pdf

7.   Department for Manufacturing, Innovation, Trade, Resources and Energy. Petroleum: Administrative Arrangements.http://www.petroleum.dmitre.sa.gov.au/environment/regulation/admin_arrangements

8.   Department for Manufacturing, Innovation, Trade, Resources and Energy. Petroleum: SEO, EIR and ESA Reports.http://www.pir.sa.gov.au/petroleum/environment/register/seo,_eir_and_esa_reports

9.   Department of Sustainability, Environment, Water, Population and Communities. Environmental Assessments.http://www.environment.gov.au/epbc/assessments/

10.   M Malavazos, The South Australian Petroleum Act 2000-principles and philosophy of best practice regulation. MESA Journal April 200121

11.   Department for Manufacturing, Innovation, Trade, Resources and Energy. Petroleum and Geothermal Energy Act Compliance Policy;   2012https://sarigbasis.pir.sa.gov.au/WebtopEw/ws/samref/sarig1/image/DDD/RB201000013.pdf

12.   Department for Manufacturing, Innovation, Trade, Resources and Energy. Petroleum: Petroleum and Geothermal Energy Act Annual

Compliance   Reporting.   http://www.pir.sa.gov.au/petroleum/
legislation/compliance/petroleum_act_annual_compliance_
report

13.   Department for Manufacturing, Innovation, Trade, Resources
and Energy. Petroleum: Annual Reports.www.pir.sa.gov.au/
petroleum/legislation/company_annual_reports

14.   International Energy Agency Golden Rules for a Golden Age of Gas,
World Energy Outlook Special Report on Unconventional Gas;
2012www.worldenergyoutlook.org/media/weowebsite/2012/
goldenrules/WEO2012_ GoldenRulesReport.pdf

15.   Society of Petroleum Engineers (SPE)Petroleum Resources
Management System; 2007

16.   DMITRE Roadmap for Unconventional Gas Projects in South
Australia, Appendix 1; 2012http://www.misa.net.au/_data/
assets/pdf_file/0009/178344/Appendix_1_Rpundtable_
Members.pdf

Key Factors for Assessing Climate Benefits of Natural Gas Versus Coal Electricity Generation

Xiaochun Zhang[1], Nathan P Myhrvold[2], and Ken Caldeira[1]

[1]Department of Global Ecology, Carnegie Institution for Science, Stanford, CA 94305, USA

[2]Intellectual Ventures, Bellevue, WA 98005, USA

ABSTRACT

Assessing potential climate effects of natural gas versus coal electricity generation is complicated by the large number of factors reported in life cycle assessment studies, compounded by the large number of proposed climate metrics. Thus, there is a need to identify the key factors affecting the climate effects of natural gas versus coal electricity production, and to present these climate effects in as clear and transparent a way as possible. Here, we identify power plant efficiencies and methane leakage rates as the factors that explain most of the variance in greenhouse gas emissions by natural gas and coal power plants. Thus, we focus on the role of these factors in determining the relative merits of natural gas versus coal power plants. We develop a simple model estimating CO_2 and CH_4 emissions from natural gas and coal power plants, and resulting temperature change. Simple underlying physical changes can be obscured by abstract evaluation metrics, thus we focus our analysis on the time evolution of global mean temperature. We find that, during the period of plant operation, if there is substantial methane leakage, natural gas plants can produce greater near-term warming than coal plants with the same power output. However, if methane leakage rates are low and power plant efficiency is high, natural gas plants can produce some reduction in near-term warming. In the long term, natural gas power plants produce less warming than would occur with coal power plants. However, without carbon capture and storage natural gas power plants cannot achieve the deep reductions that would be required to avoid substantial contribution to additional global warming.

INTRODUCTION

The most severe impacts of climate change may be avoided if efforts are made to transform the global energy systems into one that does not rely so heavily on disposing of greenhouse gases (GHGs) in the atmosphere. A transition from a global system of high GHG emission electricity generation to low GHG emission electricity generation will be central to any effort to mitigate climate change (Hoffert et al 1998, 2002, Caldeira et al 2003, Myhrvold and Caldeira 2012, WeijerMars et al 2013).

Natural gas is increasingly seen as a 'bridge fuel' for transitions to renewable and/or near-zero emission energy sources (Moniz et al 2011, Paltsev et al 2011). However, recent research has led to differing conclusions about the climate implications of more energy reliance on natural gas, and the impacts of natural gas utilization on climate are being debated (Howarth et al 2011, Wigley2011, Alvarez et al 2012, Burnham et al 2012, Grubert et al 2012, Levi 2013, Jackson et al

2014, Shearer et al 2014). Several studies have surveyed the impacts of GHG from natural gas production and utilization, focusing on different factors and leading to different conclusions (Jiang et al 2011, Karion et al 2013, Allen et al 2013, Miller et al 2013, Brandt et al 2014, Howarth 2014). On one hand, the use of natural gas emits less CO_2 per unit energy than does coal. On the other hand, because the climate is far more sensitive to methane than to carbon dioxide (Shindell et al 2012), researches have been emerging that potential climate benefits of natural gas use may be offset by leaks at gas drilling fields or other earlier points in the natural gas production lifecycle (Howarth et al 2011, Wigley 2011, Burnham et al 2012, Jackson et al 2014).

Various studies use and/or advocate for different metrics to estimate the climate impact of natural gas utilization. Studies by Howarth et al (2011), Howarth (2014) compared the emissions of natural gas, coal and other fossil fuels, and suggested that both a 20-year time horizon Global Warming Potential (GWP20) and a 100-year Global Warming Potential (GWP100) should be used when analyzing the impacts of these fuels. Alvarez et al (Alvarez et al 2012) extended the Global Warming Potential (GWP) metrices by use of integrated radiative forcing (RF) to estimate methane emission from natural gas, and found that using natural gas instead of coal for power plants can immediately reduce RF and that reducing methane emission would produce greater benefits. Burnham et al (2012) developed distribution functions to estimate methane emissions with GWP20 and GWP100 for life-cycle GHG emissions. Caldeira and Myhrvold (Caldeira and Myhrvold 2012) looked at the projected time evolution of temperature change from various power plants to conclude that, compared with the most efficient coal power plants, natural gas can at best achieve modest climate benefits on the century time scale.

In this study, we develop a power plant GHG emission model and apply available life-cycle parameters to calculate associated CO_2 and CH_4 emissions. We then feed these emissions into a simple climate model to predict resulting time evolution of RF and global mean temperature. These results make it clear that hundreds of years after a plant ceases operation, CO_2 emissions from the plant are primarily responsible for warming (indicating that on these time scales, efficiency and fuel type (natural gas versus coal) are the most important factors). Nevertheless, if natural gas leakage rates are high, CH_4 emissions can be the dominant warming effect during plant operation. Thus, controlling natural gas leakage remains a critical issue. Nevertheless, in the absence of carbon capture and storage, even in the most optimistic of scenarios, the operation of natural gas plants results in large amounts of GHGs emitted to the atmosphere. In the absence of carbon capture and storage, natural gas plants cannot produce the deep reductions in emissions that would be required of energy systems that do not contribute substantially to global warming.

METHODS

To estimate the amount of global warming that would be produced by different natural gas and coal power plants, we developed a power plant GHG emissions model, and used a schematic climate model (Myhrvold and Caldeira 2012, 2013) to investigate the RF and global mean temperature changes (ΔT). This model (see section 2.2), considers the pathway from emissions of GHGs to global mean temperature response.

Power Plant Emissions

The plants in this study are single natural gas power plants and single coal power plants with capacity of 1 GW. The major emission period of power plants is the operation period, construction emissions are relatively small in total life-cycle emissions, and relatively similar for the two types of plants (Pacca and Horvath 2002, Weisser 2007, Myhrvold and Caldeira 2012, Whitaker et al 2012, O'Donoughue et al 2014, Heath et al 2014), so construction emission is ignored in this study. The major emissions from electricity generated with natural gas are CO_2 and CH_4 (see figures 1 and SS1), although some cases considered second important emissions such as black carbon and SO_2 (Spath et al 1999, Hayhoe et al 2002, Wigley 2011). In our analysis, we do not consider climate effects of these emissions because (1) while the individual contributions of black carbon and SO_2 are large, the net climate effect of black carbon and SO_2 emissions is small (see figure 1), (2) we assume that there is motivation to reduce these emissions for non-climate (i.e. health) reasons, and (3) they are not intrinsic to the technologies being considered and may be reduced through point source pollution controls.

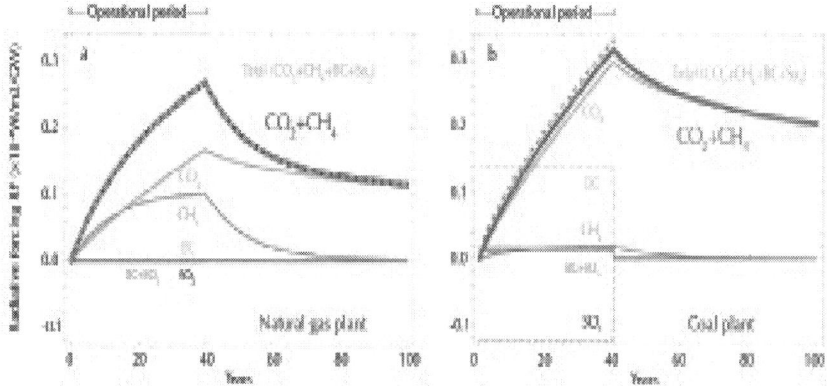

Figure 1: Attribution analysis for climate effects of (a) natural gas and (b) coal power plants, using radiative forcing (RF) as climate metric. The greenhouse gases and aerosols considered include CO_2, CH_4, black carbon (BC) and SO_2. The emissions use the relationship of the gases and aerosols emissions noted by Hayhoe et al (2002) and Wigley (2011). At the end of operational period, the net contributions to radiative forcing are dominated by greenhouse gases (CO_2 and CH_4 comprise ~100% of natural gas plant and 95.6% of coal plant radiative forcing). Individual contributions of black carbon and SO_2 are large but the combined net contributions of these aerosols are small (BC and SO_2 comprise ~0.1% of natural gas plant and 4.4% of coal power plant radiative forcing).

Annual GHG emissions were calculated by simple models.

For natural gas power plants:

Annual CO_2 emission is

$$E_{ng.CO_2} = \left[\frac{\text{molpct}_{C/ng} - R_{leak} \times \text{molpct}_{CH4/ng}}{1 - R_{leak}} \right]$$
$$\times \left[\frac{\text{Molmass}_{CO_2}}{\text{Molmass}_{ng}} \right]$$
$$\times \left[\frac{\text{Electr}_{ng}}{HV_{ng} \times \eta_{ng}} \right];$$

(1)

Annual CH_4 emission is

$$E_{ng.CH_4} = \left[\frac{R_{leak} \times molpct_{CH_4/ng}}{1 - R_{leak}} \right]$$
$$\times \left[\frac{Molmass_{CH_4}}{Molmass_{ng}} \right]$$
$$\times \left[\frac{Electr_{ng}}{HV_{ng} \times \eta_{ng}} \right].$$

(2)

For coal power plants: annual CO_2 emission is

$$E_{coal.CO_2} = masspct_{C/coal}$$
$$\times \left[\frac{Molmass_{CO_2}}{Molmass_C} \right]$$
$$\times \left[\frac{Electr_{coal}}{HV_{coal} \times \eta_{coal}} \right];$$

(3)

Annual CH_4 emission is

$$E_{coal.CH_4} = rate_{CH_4/CO_2 \cdot coal} \times E_{coal.CO_2},$$

(4)

where $molpct_{C/ng}$ is molar carbon per molar natural gas; $masspct_{C/coal}$ is the mass percent of carbon in coal; R_{leak} is natural gas leakage rate; $rate_{CH_4/CO_2 \cdot coal}$ is the ratio of CH_4 emissions to CO_2 emissions from coal mining; $molpct_{CH_4/ng}$ is the molar fraction of methane in natural gas; $Molmass_{CO_2}$, $Molmass_{CH_4}$, $Molmass_C$ and $Molmass_{ng}$ are molar masses of CO_2, CH_4, carbon and natural gas; $Electr_{coal}$ and $Electr_{coal}$ are 1 GW; HV_{ng} and HV_{coal} are heating value of natural gas and coal; η_{ng} and η_{coal} are the efficiencies of natural gas and coal power plants.

Thermal efficiency and emission rate data were taken from the National Renewable Energy Laboratory (NREL) Life Cycle Assessment Database (Whitaker et al 2012, O'Donoughue et al 2014) and recent literature (Brandt et al 2014, Howarth 2014, UNFCCC 2014). The coal power plant thermal efficiencies in this database range from 23% to 51% (Whitaker et al 2012), and the value for a world typical coal plant is 34% (WEC 2013), with 4.18‰ methane leakage (the ratio of methane emission to carbon emission from coal use). It should be noted that the 51% efficiency is reported for an experimental black coal integrated gas combined cycle (IGCC) system (May and Brennan 2003) and may exceed what is feasible for commercial operation. The natural gas power plant thermal efficiencies range from 25% to 60% (O'Donoughue et al 2014), and the value for a typical natural gas power plant is 40% (WEC 2013). The methane leakage rates (the ratio of methane emission to total natural gas use) range from 0% to 9% (Brandt et al 2014, Howarth 2014). For more details of parameter values please see S2, and relative literatures (McGurlet al 2004, EPA 2008, Whitaker et al 2012, Zhang et al 2012a, 2012b, 2013, O'Donoughue et al 2014, Brandt et al 2014, Howarth2014)

An important aim of this paper is to identify a small handful of key factors that explain most of the difference in climate effects between natural gas and coal based power production. The model for natural gas power plant GHG emissions represented by equations (1) and (2) considers thermal efficiency and natural gas leakage rates, and explains 98.1% of the variance among natural gas power plant GHG emissions as determined by the harmonization study of O'Donoughue et al (2014). The next most important factor explaining variance in these emission rates is CO_2 emission during natural gas production and transportation. The model for coal power plant GHG emissions represented by equations (3) and (4) considers thermal efficiency and coal mine methane emissions, and explains 98.2% of the variance in coal power plant GHG emissions as determined by the harmonization study of Whitaker et al (2012). The model presented here is intended to provide a quantitative and conceptual understanding of the role of key factors affecting all natural gas and coal based electricity generation. In particular cases, such as with liquid natural gas fueled power plants, where transportation emissions could be high, other factors could prove important. In these cases, it may be important to specify the full range of parameters in equations (1) through (4).

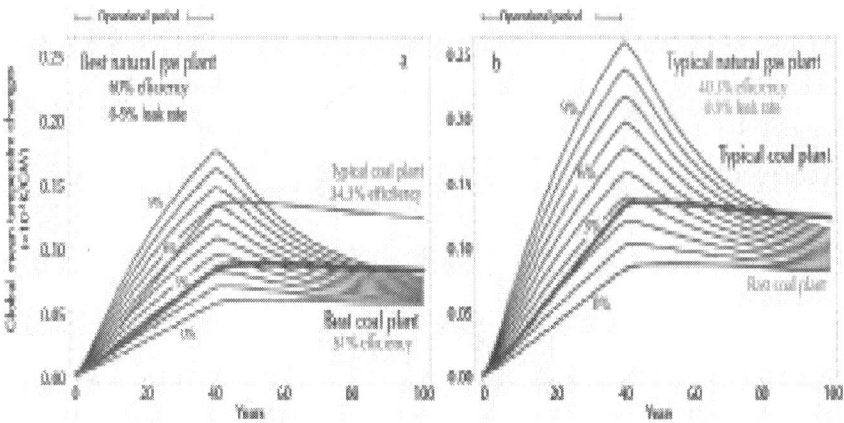

Figure 2: Global temperature change from the (a) most efficient and (b) typical efficiency natural gas and coal power plants. The operational period is 40 years. Efficiencies of natural gas and coal power plants are 60% and 51% for the most efficient natural gas and coal power plants, respectively, (Whitaker et al 2012, O'Donoughue et al 2014); 40.3% and 34.3%, respectively, for global fleet average efficiencies (world typical plants, WEC 2013). The range of natural gas leakage rate considered is 0%–9% (Howarth 2014). Amount of warming from natural gas versus coal power plants for the first several decades depends sensitively on natural gas leakage rates and power plant efficiencies. Several decades after the plant ceases to operate, most of the methane no longer resides in the atmosphere; at that time, the relative warming from natural gas versus coal power plants depends primarily on power plant efficiency.

Climate Model

GHG Residence in the Atmosphere

Emissions of GHGs were converted to concentrations of GHGs in the atmospheric environment. The amount of GHGs in the atmosphere is:

$$\text{mass}_{\text{GHG}}(t) = \int_0^{t_m} E_{\text{GHG}}(t) \cdot G(t_m - t)dt,$$

(5)

Where $E_{\text{GHG}}(s)$ is annual GHG emission, $0 \leq t \leq t_m$.

In this study, we applied Joos et al's (Joos et al 2013) parameter of CO_2 impulse response function and Prather et al's (Prather et al 2012) parameters of CH_4 impulse response function to estimate atmospheric GHG concentrations

$$G_{CO_2}(t) = 0.2173 + 0.2240\, e^{(-t/394.4)}$$
$$+ 0.2824\, e^{(-t/36.54)} + 0.2763\, e^{(-t/4.304)}, \tag{6}$$

$$G_{CH_4}(t) = e^{(-t/12.4)}, \tag{7}$$

$$G_{CO_2 \leftarrow CH_4}(t) = 0.5977 + 0.6361\, e^{(-t/394.4)}$$
$$+ 1.1758\, e^{(-t/36.54)}$$
$$- 2.0056\, e^{(-t/12.4)} - 0.4040\, e^{(-t/4.304)}. \tag{8}$$

Radiative Forcing (RF)

The RF was calculated using equations provided by the Intergovernmental Panel on Climate Change (IPCC, 2013).

$$RF\left(mass_{GHG}(t)\right) = RF_{CO_2}\left(mass_{CO_2}(t)\right)$$
$$+ RF_{CH_4}\left(mass_{CH_4}(t)\right)$$
$$+ RF_{CO_2}\left(mass_{CO_2 \leftarrow CH_4}(t)\right), \tag{9}$$

Where

$$RF_{CO_2}\left(mass_{CO_2}(t)\right) = 3.35\Big[g\left(cct_{bs.CO_2}\right.$$
$$+\ \alpha_{CO_2}\cdot mass_{CO_2}(t)\big)$$
$$-\ g\left(cct_{bs.CO_2}\right)\Big],$$

(10)

$$RF_{CH_4}\left(mass_{CH_4}(t)\right) = 0.036\Big[\left(cct_{bs.CH_4}\right.$$
$$+\ \alpha_{CH_4}\cdot mass_{CH_4}(t)\big)^{1/2}$$
$$-\ \left(cct_{bs.CO_2}\right)^{1/2}\Big] - f_{CH_4},$$

(11)

$$g\left(cct_{CO_2}\right) = Log\left(1+1.2\ cct_{CO_2} + 0.005\ cct_{CO_2}^2 \right.$$
$$+\ 1.4\times 10^{-6}cct_{CO_2}^3\big)$$

(12)

And CO_2, CH_4 are constants used to convert the mass to concentrations cct is concentration of GHG in the atmosphere. $cct_{bs.CO_2}$ and $cct_{bs.CH_4}$ are the baseline concentrations of CO_2 and CH_4 in the atmosphere before emissions. f_{CH_4} is a function about concentrations of CH_4 and N_2O, and here f_{CH_4} is zero. For more details please see SE1.3 of (Myhrvold and Caldeira 2012).

Temperature Change

Global mean temperature changes (ΔT) were estimated from the RF by using a simple energy balance model that represents the effective heat capacity of the climate system as a one-dimensional diffusive column, as described in (Caldeira and Myhrvold 2013):

$$\frac{\partial \Delta T}{\partial t} = k_v \frac{\partial^2 \Delta T}{\partial z^2},$$

(13)

$$\left.\frac{\partial \Delta T}{\partial z}\right|_{z=0} = \left.\frac{\lambda \Delta T - \mathrm{RF}(t)}{\rho \, k_v \, f \, c_p}\right|_{z=0},$$

(14)

$$\Delta T \big|_{t=0} = 0,$$

(15)

$$\left.\frac{\partial \Delta T}{\partial z}\right|_{z=z_{max}} = 0.$$

(16)

Model parameters were chosen to mimic median results from the Climate Model Intercomparison Project phase 5 (CMIP5, Caldeira and Myhrvold 2013). The climate sensitivity parameter (λ) is 1.051. The ratio of adjusted RF to the classical RF derived from the IPCC formula is 0.775. The thermal diffusivity (k_v) is 4.24×10^3 m^2 s^{-1}. For more details, see (Myhrvold and Caldeira2012, 2013).

Scenarios

To project the amount of global warming that would be caused by the operation of different power plants, we performed simulations using the climate model described above for each natural gas plant described in O'Donoughue et al (2014) and each coal plant described in Whitaker et al (2012). These plants were assumed to operate for 40 years.

To quantify the relative importance of methane leakage rates versus variation in power plant efficiency, we analyzed four end-member scenarios for natural gas power plants. The scenarios are high GHG emission with high methane leakage rate and low thermal efficiency, low GHG emission with zero methane leakage rate and high thermal efficiency, zero methane leakage rate and low thermal efficiency, and high methane leakage rate and high thermal efficiency. To compare, there are also three scenarios for coal power plants which are low, high and typical thermal efficiencies. All high and low values were taken from the National Renewable Energy Laboratory (NREL) natural gas and coal LCA harmonization studies (Whitaker et al 2012, O'Donoughue et al 2014). These simulations were done for two scenarios: one in which the plants operated for a 40 year period and another in which the plants were assumed to continue operating indefinitely.

RESULTS

Simulations Based on Full Set of Natural Gas and Coal Power Plant Lca Studies

We used our simple climate model to project changes in atmospheric GHG concentrations, RF, and temperature change using LCA data. Results for the full set of power plants in the NREL natural gas and coal LCA studies (Whitaker et al 2012, O'Donoughueet al 2014) and recent literature (Brandt et al 2014, Howarth 2014) can be found in figures 2 and S2. In these figures, global mean temperatures increase for the first 40 years, while the power plants are in operation. The power plants are turned off at year 40. Temperatures for the coal plants and the least-warming natural gas plant remain relatively constant even after emissions cease. Temperatures decrease over a period of decades after the cessation of the natural gas power plants that produce the most warming. This is a result of the chemical breakdown of methane in the atmosphere and ultimately the radiation of energy to space. The part of warming that is caused by CO_2 releases from the natural gas power plant persists far longer, as is the case with CO_2 emissions from coal power plants. However, the range of temperature increases for natural gas plants span the range of increases for coal plants for most of the century.

To quantify the extent to which variation in methane leakage rates versus variation in power plant efficiency explained the range of results seen in figures 2 and S2, we show the highest efficiency gas plant with the highest methane leakage (9%) and the lowest efficiency gas plant with no methane leakage. The most efficient natural gas plant (60%) even with the highest leakage rate (9%) produced less warming than the coal plant at the end of the century. However, if integrated measures are used that take into consideration average changes over a century, the most efficient coal plants would be considered better than the most efficient natural gas plant with such high upstream leakage rates.

Importance of Methane Leakage and Plant Efficiency

To examine the climate consequences of different methane leakage rates, we performed a series of simulations (figures 2, 3 and S4) using methane leakage rates in the range of 0%–9% (Brandt et al 2014, Howarth 2014) using both world typical (WEC 2013) and best (Whitaker et al 2012, O'Donoughue et al 2014) natural gas power plant efficiencies. For the 40-year period of plant operation, in the case of the most efficient (60%) power plant, a 2% leakage rate maintains approximate warming parity with the most efficient (51%) coal plant (figure 2(a)). However, during this 40-year operational period, the most efficient gas plant maintains approximate parity with a typical efficient (34.3%) coal plant with methane leakage rates of 5 or 6%. By the end of the century, 60 years after the cessation of power generation, warming from the coal plants considerably exceed the amount of warming from natural gas plants, even with 9% methane leakage. By the end of the century, the most efficient natural gas plant is producing 6.3%–35.0% less warming than the most efficient coal plant, and 40.0%–58.4% less warming than the typical coal plant, depending on the methane leakage rate in the 0%–9% range. For the several centuries period of plant operation, at the end of 1000 years, the most efficient natural gas power plant is producing 12.5%–31.3% less warming than the most efficient coal plant, and 41.7%–54.2% less warming than the typical coal plant, depending on the methane leakage rate in the 0%–9% range.

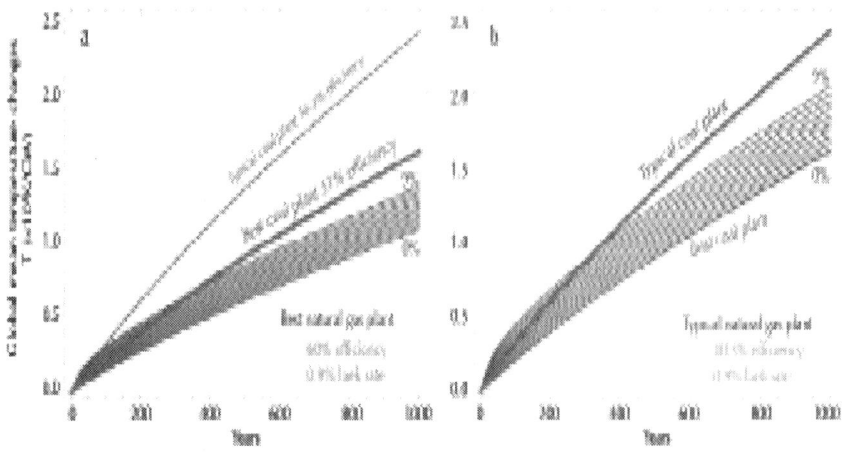

Figure 3: Global temperature change from continuous operation of the (a) most efficient and (b) typical efficiency natural gas and coal power plants. Efficiencies of natural gas and coal power plants are 60% and 51% for the most efficient plants; 40.3% and 34.3% for global fleet average efficiencies (world typical plants). Range of natural gas leakage rates is from 0% to 9%. Within this range of leakage, after several centuries, the best natural gas plants produce less warming than coal plants. In the long term, power plant efficiency (and whether the plant is burning natural gas or coal) is more important than the rate of methane leakage.

The importance of methane leakage rate relative to power-plant efficiency diminishes when longer time horizons are considered. To illustrate fundamental physical characteristics of GHGs and the climate system, we simulated 1000 years of continuous operation (figure 3), using the same power plants as in figure 2, to show that the relative importance of methane leakage rate relative to efficiency diminishes when longer time horizons are considered. It should not be seen as a realistic projection of future resource use. At year 100, the most efficient natural gas plant warms from 32% less than to 41% more than the most efficient coal plant, depending on methane leakage in the 0%–9% range. By year 1000, the most efficient natural gas plant avoids from 12.5% to 31.3% of the warming that would be produced by the most efficient coal plant. Again, we note that the results reported for the most efficient coal plant used here are an experimental plant. No plant in commercial operation has attained such efficiencies.

DISCUSSION

The results presented above show that whether natural gas plants are better than coal plants cannot be answered in the general case. In terms of direct climate effects, the question can be answered with reference to a particular natural gas plant with a particular upstream methane leakage rate when compared with a particular coal plant using a specified metric evaluated over a specified time interval. In the following discussion, we use 'better' and 'worse' to describe whether one power plant would produce less or more, respectively, of a temperature change than another power plant.

We recognize that climate effects would be only part of a more comprehensive power plant evaluation. Natural gas plants differ from coal plants not only in their direct climate consequences, but also in other important ways. For example, natural gas plants can provide reserve services to the power grid, while coal plants can do this to a much lesser degree (Armaroli and Balzani 2011). Therefore, natural gas plants might facilitate near-term integration of larger percentage of renewables, assuming policy drivers for deployment of renewables were in place (Richardson 2013). Further, natural gas can offer significant benefits over coal plants in terms of air quality (Han et al 2009, Zhan et al 2009, Zhang et al 2009, de Gouw et al 2014). Evaluating the relative benefits of these characteristics of gas and coal plants is beyond the scope of our study. Nevertheless, with respect to climate benefits of natural gas versus coal power plants, some general results have emerged from our study.

In this study, we assessed and discussed key factors affecting the evaluation of the climate consequences of natural gas versus coal electricity generation. The key factors affecting the climate consequences of natural gas plants with the same power output are the efficiency of the gas plant and the leakage of methane associated with natural gas supply. The key factor affecting climate consequences of coal plants (assuming appropriate scrubbing of SO_2 and black carbon) is the power plant efficiency. Other factors such as leakage of methane from coal beds could also potentially be important in some circumstances. Thus, the evaluation from a climate perspective of the relative merits of a natural gas plant and a coal plant largely depend on these power plant efficiencies and methane leakage. In general, because natural gas produces more thermal energy per unit carbon and natural gas plants can be more efficient than coal plants, natural gas power plants produce less warming than coal plants in the long term (figure 3). However, because methane is a potent but relatively short-lived GHG, if there is substantial methane leakage associated with natural gas supply, then natural gas plants can produce more warming than a comparable coal plant during the period of operation.

Whether a natural gas plant with a certain methane leakage rate and efficiency is better or worse than a coal plant, depends not only on the metric being considered but also on the characteristics of the reference coal plant. Natural gas power plants need lower methane leakage rates or higher efficiencies to perform better than higher efficiency coal plants (figures 2, S4 and S5). Some of the apparent disagreement among studies was due to the fact that different studies assumed different reference coal power plants (Howarth et al 2011, Myhrvold and Caldeira 2012, Howarth 2014).

Different reference power plants are relevant to different situations. For example, if someone were considering building a new power plant today, they could decide between building a high efficiency natural gas plant (60% efficiency) versus a high efficiency coal plant (51% efficiency) (figure 2(a)). In this case, if the methane leakage rate is above about 2%, then the natural gas plant will produce more warming in the first decades than will the coal plant. It should be noted that power plants with 51% efficiency are not commercially available today (May and Brennan 2003), but this could represent an upper bound efficiency for some future commercial coal plant. However, for plants with a 40-year operation time, the natural gas plant will produce less warming at 100 years, even with a 9% methane leakage rate. If someone were considering whether to replace a typical coal plant (34.3% efficiency) with the best natural gas plant (60% efficiency), the natural gas plant would cause less climate change throughout even with a 4% or 5% methane leakage rate.

Furthermore, if someone were faced with the alternative of shutting down a typical gas plant versus a typical coal plant, there would be climate benefit to shutting down the gas plant in the near term if methane leakage was below about 2%; however, there would be long-term climate benefit even if methane leakage rates were substantially greater (figures 2(b) and S5(c)).

CONCLUSIONS

This study has focused on comparison of natural gas power plants versus coal power plants and has emphasized the climate effects of electricity generation, specifying conditions when one type of plant becomes better than another. However, it is important to note that in many cases when a natural gas plant is deemed 'better' than a reference coal plant, it is only somewhat better than the reference coal plant (figures 2 and 3). For example, the best natural gas plant with a zero methane leakage rate still produces about two-thirds of the century-integrated warming as does the best coal power plant (figure 2(a)). High-efficiency natural gas plants with low methane leakage rates can, in principle, produce half the century-integrated warming as today's typical coal plant. Thus, there is potential climate benefit in replacing low-efficiency coal plants with high-efficiency, low-methane leakage natural gas plants.

However, many well-publicized GHG emission targets require much deeper cuts in emissions than can be provided by natural gas. For example, California's AB32 regulation calls for 80% reductions in emissions below 1990 levels by mid-century (CARB2007). Many power plants built today could still be operational in mid-century; this raises the question of the extent to which natural gas can help in achieving these policy objectives. If natural gas is to play a long-term role in electricity production in a world with greatly constrained carbon emissions, then carbon capture and storage may be an essential component of future natural gas systems (Davidson et al 2005).

Natural gas is thought of as a 'bridge' fuel by some policy makers—a temporary fuel to be used until a transition to near-zero emission technologies becomes more feasible (Moniz et al 2011, Paltsev et al 2011, BP 2014). Thus, natural gas is promoted as a way to decrease near-term emissions as we make a transition to energy systems that deeply cut long-term emissions. However, if methane leakage rates cannot be maintained at very low values, near-term climate benefits may be small or non-existent. There is potential that, relative to coal, the deployment of natural gas power plants could both produce excess near-term warming (if methane leakage rates are high) and produce excess long-term warming (if the deployment of natural gas plants today delays the transition to near-zero emission technologies. Thus, achieving climate benefits from the use of natural gas depends on building high-efficiency natural gas plants, controlling methane leakage, and on developing a policy environment that assures a transition to future lower-emission technologies.

ACKNOWLEDGEMENTS

We thank Ms Dawn Ross (Carnegie Institution for Science) and Dr Dan Freeman (Intellectual Ventures) for help in preparing this document. We thank the anonymous referees for their constructive comments.

REFERENCES

1. Allen D T et al 2013 Measurements of methane emissions at natural gas production sites in the United States Proc. Natl Acad. Sci. USA 110 17768–73.

2. Alvarez R A, Pacala S W, Winebrake J J, Chameides W L and Hamburg S P 2012 Greater focus needed on methane leakage from natural gas infrastructure Proc. Natl Acad. Sci. USA 109 6435–40.

3. Armaroli N and Balzani V 2011 Energy for a sustainable world (New York: Wiley) Online: (www.isof.cnr.it/sites/default/files/users/armaroli/Wiley_2011.pdf).

4. Brandt A R et al 2014 Methane leaks from North American natural gas systems Science 343 733–5

5. Burnham A, Han J, Clark C E, Wang M, Dunn J B and Palou-Rivera I 2012 Life-cycle greenhouse gas emissions of shale gas, natural gas, coal, and petroleum Environ. Sci. Technol. 46 619–27

6. Caldeira K, Jain A K and Hoffert M I 2003 Climate sensitivity uncertainty and the need for energy without CO2 emissionScience 299 2052–4

7. Caldeira K and Myhrvold N P 2012 Temperature change versus cumulative radiative forcing as metrics for evaluating climate consequences of energy system choices Proc. Natl Acad. Sci. USA 109 E1813–1813

8. Caldeira K and Myhrvold N P 2013 Projections of the pace of warming following an abrupt increase in atmospheric carbon

dioxide concentration Environ. Res. Lett. 8 034039

9. CARB 2007 California Air Resources Board. Assembly Bill 32: Global Warming Solutions Act (www.arb.ca.gov/c.c./ab32/ab32.ht)

10. Chou V, Keairns D, Turner M, Woods M and Zoelle A 2014 Quality guidelines for energy system studies: process modeling design parameters (National Energy Technology Laboratory) pp 13–21 NETL/DOE-341/051314

11. de Gouw J A, Parrish D D, Frost G J and Trainer M 2014 Reduced emissions of CO2, NOx, and SO2 from US power plants owing to switch from coal to natural gas with combined cycle technology Earths Future 2 75–82

12. Dones R, Bauer C, Bolliger R, Burger B, Faist Emmenegger M, Frischknecht R, Heck T, Jungbluth N, Röder A and Tuchschmid M 2007 Life cycle inventories of energy systems: results for current systems in Switzerland and other UCTE countries Ecoinvent Rep. 5   Online:(www.ecolo.org/documents/documents_in_english/Life-cycle-analysis-PSI-05.pdf)

13. EPA 2008 The Climate Leaders Greenhouse Gas Inventory Protocol Design Principles Guidance: Stationary Combustion Sources (Washington DC: US Environmental Protection Agency)

14. Grubert E A, Beach F C and Webber M E 2012 Can switching fuels save water? A life cycle quantification of freshwater consumption for Texas coal- and natural gas-fired electricity Environ. Res. Lett. 7 045801

15. Han G, Zhan S, Zhang X and Ma C 2009 Influence factors and mathematical model of coal dust particles threshold velocity J. China Coal Soc. 10 013

16. Hayhoe K, Kheshgi H S, Jain A K and Wuebbles D J 2002 Substitution of natural gas for coal: climatic effects of utility sector emissions Clim. Change 54 107–39

17. Heath G A, O'Donoughue P, Arent D J and Bazilian M 2014 Harmonization of initial estimates of shale gas life cycle greenhouse gas emissions for electric power generation Proc. Natl Acad. Sci. USA 111 E3167–76

18. Hoffert M I et al 1998 Energy implications of future stabilization of atmospheric CO2 content Nature 395 881–4

19. Hoffert M I et al 2002 Advanced technology paths to global climate stability: energy for a greenhouse planet Science 298981–7

20. Howarth R W 2014 A bridge to nowhere: methane emissions and the greenhouse gas footprint of natural gas Energy Sci. Eng. 1 1–14

21. Howarth R W, Santoro R and Ingraffea A 2011 Methane and the greenhouse-gas footprint of natural gas from shale formations Clim. Change 106 679–90

22. IPCC 2013 Climate Change 2013: The Physical Science Basis

23. Jackson R B, Down A, Phillips N G, Ackley R C, Cook C W, Plata D L and Zhao K 2014 Natural gas pipeline leaks across Washington, DC Environ. Sci. Technol. 48 2051–8

24. Jiang M, Griffin W M, Hendrickson C, Jaramillo P, VanBriesen J and Venkatesh A 2011 Life cycle greenhouse gas emissions of Marcellus shale gas Environ. Res. Lett. 6 034014

25. Joos F et al 2013 Carbon dioxide and climate impulse response functions for the computation of greenhouse gas metrics: a multi-model analysis Atmos. Chem. Phys. 13 2793–825

26. Karion A et al 2013 Methane emissions estimate from airborne measurements over a Western United States natural gas field Geophys. Res. Lett. 40 4393–7

27. Levi M 2013 Climate consequences of natural gas as a bridge fuel Clim. Change 118 609–

28. May J R and Brennan D J 2003 Life cycle assessment of Australian fossil energy options Process Saf. Environ. Prot. 81317–30

29. Metz B, Davidson O, de Coninck H C, Loos M and Meyer L A 2005 IPCC Special Report on Carbon Dioxide Capture and Storage. Prepared by Working Group III of the Intergovernmental Panel on Climate Change (Cambridge: Cambridge University Press)

30. Miller S M et al 2013 Anthropogenic emissions of methane in the United States Proc. Natl. Acad. Sci. USA 110 20018–22

31. Moniz E J, Jacoby H D, Meggs A J M, Armtrong R C, Cohn D R, Connors S R, Deutch J M, Ejaz Q J, Hezir J S and Kaufman G M 2011 The future of natural gas Camb. MA Mass. Inst. Technol. Online: (http://mydocs.epri.com/docs/summerseminar11/presentations/03-02_moniz_mit_natural_gas_v1.pdf)

32. Myhrvold N P and Caldeira K 2012 Greenhouse gases, climate change and the transition from coal to low-carbon electricity Environ. Res. Lett. 7 014019

33. O'Donoughue P R, Heath G A, Dolan S L and Vorum M 2014 Life cycle greenhouse gas emissions of electricity generated from conventionally produced natural gas J. Ind. Ecol. 18 125–44

34. Pacca S and Horvath A 2002 Greenhouse gas emissions from building and operating electric power plants in the Upper Colorado River Basin Environ. Sci. Technol. 36 3194–200

35. Paltsev S, Jacoby H D, Reilly J M, Ejaz Q J, Morris J, O'Sullivan F, Rausch S, Winchester N and Kragha O 2011 The future of US natural gas production, use, and trade Energy Policy 39 5309–21

36. Prather M J, Holmes C D and Hsu J 2012 Reactive greenhouse gas scenarios: systematic exploration of uncertainties and the role of atmospheric chemistry Geophys. Res. Lett. 39 L09803

37. Richardson D B 2013 Electric vehicles and the electric grid: a review of modeling approaches, Impacts, and renewable energy integration Renew. Sustain. Energy Rev. 19 247–54

38. Shearer C, Bistline J, Inman M and Davis S J 2014 The effect of natural gas supply on US renewable energy and CO2 emissions Environ. Res. Lett. 9 094008

39. Shindell D et al 2012 Simultaneously mitigating near-term climate change and improving human health and food security Science 335 183–9

40. Spath P L, Mann M K and Kerr D R 1999 Life Cycle Assessment of Coal-Fired Power Production (Golden, CO (US): National Renewable Energy Laboratory) Online: (www.osti.gov/scitech/biblio/12100)

41. UNFCCC 2014 United Nations Framework Convention on Climate Change: 2014 Annex I Party GHG Inventory Report (Geneva: United Nations Office at Geneva) Online: (www.unfccc.de/index.html)

42. WEC 2013 Energy Efficiency Indicators database (World Energy Council) Online: (www.worldenergy.org/data/efficiency-indicators)

43. WeijerMars R et al 2013 Energy strategy research—Charter and perspectives of an emerging discipline Energy Strategy Rev. 1

135–7

44. Weisser D 2007 A guide to life-cycle greenhouse gas (GHG) emissions from electric supply technologies Energy 32 1543–59

45. Whitaker M, Heath G A, O'Donoughue P and Vorum M 2012 Life cycle greenhouse gas emissions of coal-fired electricity generation J. Ind. Ecol. 16 S53–72

46. Wigley T M L 2011 Coal to gas: the influence of methane leakage Clim. Change 108 601–8

47. Zhan S, Zhang X and Ma C 2009 Coal classification based on environmental protection and burning quality J. China Coal Soc. 34 1535–9

48. Zhang G, Zhan S and Zhang X 2009 Theory and Technology for Aerosol Pollution Control in Port (Beijing: China Communication Press)

49. Zhang X C, Chen W P, Ma C, Zhan S F and Jiao W T 2012a Regional atmospheric environment risk source identification and assessment Environ. Sci. 33 4167–72

50. Zhang X, Chen W, Ma C and Zhan S 2012b Modeling the effect of humidity on the threshold friction velocity of coal particles Atmos. Environ. 56 154

51. Zhang X, Chen W, Ma C and Zhan S 2013 modeling particulate matter emissions during mineral loading process under weak wind simulation Sci. Total Environ. 449 168–73.

Chapter 4

Electrocatalysis and Electrocatalysts for Low Temperature Fuel Cells: Fundamentals, State of the Art, Research and Development

Hartmut WendtI, Estevam V. SpinacéII, Almir Oliveira NetoII, and Marcelo LinardiII

IBerl Institut TU Darmstadt, Wendt Ingenieure, Darmstaedter Str. 65, D-64807 Dieburg, Germany

IIInstituto de Pesquisas Energéticas e Nucleares, Av. Prof. Lineu Prestes, 2242, 05508-900 São Paulo - SP, Brasil

ABSTRACT

This article deals with electrocatalysis and electrocatalysts for low temperature fuel cells and also with established means and methods in electrocatalyst research, development and characterization. The intention is to inform about the fundamentals, state of the art, research and development of noble metal electrocatalysts for fuel cells operating at low temperatures.

INTRODUCTION

The idea of electrochemical electricity generation from chemical energy in fuel cells originated already more the 150 years ago from the experiments of Grove and Schönbein[1]. However, the technical realisation had to wait for more than a hundred years as the relevant materials problems could not be solved for a very long time and the theoretical background of electrochemistry, not to speak of electrocatalysis, took so much time to be developed. It was apparent from the very beginning that the electrochemical combustion of any fuel (Equation 1) in an electrochemical cell brought about by separate anodic oxidation of a gaseous fuel, for instance hydrogen (Equation 1a), at the anode and cathodic reduction of oxygen (Equation 1b) needed the catalytic activation of the reactants, fuel as well as oxygen.

$$\text{Total reaction: fuel} + (x + y/_2)\ O_2 \rightarrow x\ CO_2 + y\ H_2O \quad (1)$$

$$\text{Anodic reaction: } H_2 + 2\ H_2O \rightarrow 2\ H_3O^+ + 2\ e^- \quad (1a)$$

$$\text{Cathodic reaction: } \tfrac{1}{2}\ O_2 + 2\ H_3O^+ + 2\ e^- \rightarrow 3\ H_2O \quad (1b)$$

Hydrogen (H_2) is a relatively stable molecule – and so are carbonaceous fuels. Due to the strength of the H-H bond or H_2-binding energy respectively of approximately 430 kJ/mol[2,3], only catalytic activation of this molecule would enhance its reaction at ambient temperature to a degree, which allows it to react with sizeable rate. For double-bonded and hence more strongly bonded oxygen (DH @ 500 kJ/mol) the catalytic activation is even more necessary. This statement holds for gas phase reactions (Equation 1) as well as for electrochemical conversion at electrode surfaces (Equation 1a and 1b). And as will be explained below, platinum in both cases is the catalyst of choice as already realized in the 19th century by Döbereiner[4]. Although this article deals almost exclusively with electrocatalysis in fuel cells, it is stressed that electrocatalysis is of high practical relevance in industrial electrochemistry. The present technology of chloroalkali electrolysis today depends essentially on electrocatalysis of the hydrogen and chlorine evolution reactions[5,6]. In the mid 80ties of the last century the phosphoric acid fuel cell technology gained shape. As documented by a number of important patents of that time it became evident that platinum and also a number of platinum alloys can be prepared on industrial scale in well-defined nanodisperse form, thus enabling fuel cell producing industries to make highest possible use of this expensive material[7-10]. In the same context also the use of active carbon supported nanodisperse platinum had been introduced into the practice of technical production of fuel cells[11].

The developments of modern fuel cells began with moderately high temperature alkaline cells[12] in the 20ies (working at 200 °C) and with high temperature molten carbonate cells[13] in the 50ties of the last century. A first technical application was the development of the alkaline cells for the Gemini space mission by United Technologies Corporation (UTC)[13,14]. The Varta-Siemens approach based on Raney-nickel electrodes in 30 wt% KOH did not lead to a convincing success[14]. Much more promising was the next step undertaken by UTC to introduce membrane fuel cells, which after disappointing experimentations with classical ion exchange membranes led to a technical brake through by introducing perfluorinated sulfonic acid membranes as so-called solid polymer electrolyte[15,16], which had been perfected by Siemens for powering submarines during the last years. Dupont initially had developed this type of membrane under the name of Nafion for the chloroalkali industry[17]. This eventually yielded into the now well-established polymer electrolyte membrane fuel cell (PEMFC) technology (Figure 1). The scanning electron microscopic cut through a so-called membrane/electrode assembly (MEA) of a PEMFC (Figure 2) shows the minute dimensions of only approximately 10 µm thick porous electrodes on both sides of the approximately 150 µm thick membrane. The membrane chemical composition is shown in Figure 3. The essential catalyst for hydrogen/oxygen fuel cells on the anode- and the cathode side is nano-dispersed platinum, supported on active carbon. Its preparation and morphology will be described below.

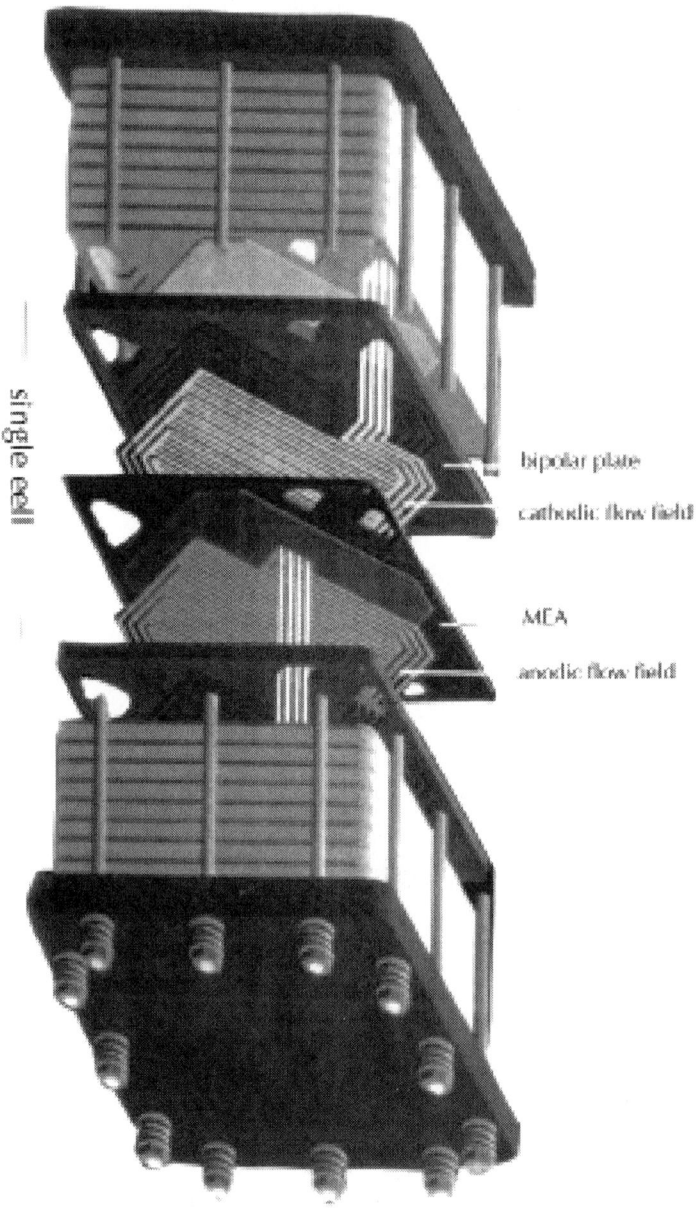

single cell

bipolar plate

cathodic flow field

MEA

anodic flow field

Figure 1: View of Proton Exchange Membrane Fuel Cell stack; included is the exploded view of a singular cell containing bipolar wall with flow field and at the centre the MEA (Membrane-Electrode-Assembly).

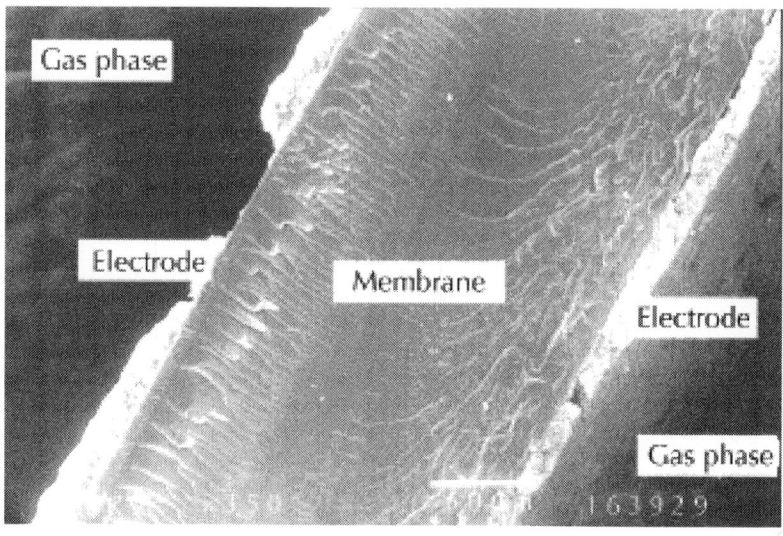

Figure 2: SEM micrograph of a typical MEA produced in the laboratory of the authors.

$$\left|-(-CF_2-CF_2-)_n-CF-CF_2-\right|_x$$
$$(O-CF_2-CF-)_m-O-CF_2-CF_2-SO_3H$$
$$CF_3$$

n - 5 to 13
x = ca. 1000
m = 1 to 3

Figure 3: Schematic presentation of the chemical structure of Nafion.

The further advance of PEMFC-electrocatalyst development was due to the fact that for mobile or for stationary electricity generation with PEMFC using hydrogen, purified hydrogen in particular, seemed to be too expensive and therefore impractical. The relatively reactive and easily processable methanol but also natural gas, which demands higher processing efforts, are cheaper than hydrogen. Methanol additionally because it is a liquid at ambient temperature and pressure is easy to store and more appropriate as fuel for cars whereas natural gas is the primary fuel of choice for stationary electricity generation. These hydrogen carriers can be converted to hydrogen by the sequence of reforming and shift-conversion (Equations. 2 and 3) that produces gas mixtures containing hydrogen and carbon dioxide but also the catalyst poison carbon monoxide. The latter is found in concentrations of up to one percent because shift conversion is an equilibrium reaction, which favours CO at higher temperature and the equilibrium concentration of CO in reformate gas is that high at the operation temperature of the heterogeneously catalysed shift conversion process (200 °C).

Reformate from Methane

Methane reforming:

$$CH_4 + H_2O \rightarrow CO + 3H_2 \ (800\text{-}900°C) \tag{2a}$$

Shift conversion:

$$CO + H_2O \rightarrow CO_2 + H_2 \ (\text{beginning at 400 and ending at 200°C}) \tag{2b}$$

Reformate from Methanol

Methanol cracking and reforming:

$$CH_3OH \rightarrow CO + 2H_2 \ (300\text{-}400°C) \tag{3a}$$

Shift conversion:

$$CO + H_2O \rightarrow CO_2 + H_2 \ (\text{beginning at 400 and ending at 200°C}) \tag{3b}$$

The generation of hydrogen from carbonaceous hydrogen carriers therefore is highly problematic with respect to the CO – sensitive platinum catalysts if the electrochemical reaction is to be performed at temperatures of between 60 and 80 °C, which is the case in PEMFCs. Their working temperature must be kept that low because of the necessary moisture in the membrane, which keeps the ionic conductivity of the membrane sufficiently high. At these low temperatures the catalytic activity of platinum is severely impaired by strongly adsorbed CO, as long as the CO concentration is not kept lower than only 10 ppm. Direct methanol fuel cells (DMFCs)[18] for automotive power generation and for so-called portables applications are even more demanding with respect to the anodic electrocatalysts. There the electrocatalyst has not only to provide for the anodic fission of O-H and C-H bonds by chemisorption of the respective fuel but the anodic oxidation of methanol ends up generating adsorbed CO in situ and thus the reaction on Pt is poisoning the catalyst. And this catalyst poison has to be removed steadily. Therefore, during the last years a couple of cocatalysts has been investigated and applied which enhance particularly the oxidation rate of CO in order to render anodic PEM-electrocatalysts more CO-tolerant. Among these cocatalysts ruthenium is the most prominent. It is technically applied since several years and it is very well documented and investigated[19]. Binary Pt/Ru catalysts supported on active carbon are commercial products. This article will therefore deal not only with dispersed platinum electrocatalysts, their preparation and characterization but will report briefly also on what we know today about composition, structure and nanomorphology of more elaborate nanodisperse, composite electrocatalyst systems. These more involved anodic catalyst systems, as far as they are used in PEMFCs, are almost exclusively supported on active carbon[20-22].

Phosphorous acid fuel cells (PAFCs) working at 200 °C are insensitive against CO and work on reformate with 1% of CO (without CO removal). But according to an announcement of UTC in 2003 (source: Internet, Hyweb) they will develop PEM fuel cells in preference to PAFCs for stationary electricity generation because the PAFC – technology does not promise any further cost reduction below the established system costs of 4,500 USD per kW.

FUNDAMENTALS: PLATINUM, THE CATALYST OF CHOICE FOR ANODIC HYDROGEN OXIDATION AND CATHODIC OXYGEN REDUCTION

Exchange Current Densities, an Indicator of Electrocatalysis

Current voltage correlations measured for the Butler-Volmer equation (Equation 4), describe mathematically the current-voltage correlation for a reversible electrochemical reaction, for instance, the anodic oxidation of hydrogen and cathodic hydrogen evolution. It comprises an anodic (*index a*) and a cathodic (*index c*) exponential terms with the overpotential h being the difference between actual electrode potential E and equilibrium potential E_o (h = E - E_o) and the charge transfer coefficients a_a and a_c in the exponent. Instead of current, I, the current density (current per unit surface), i, is chosen as the adequate quantity for this correlation, in order to obtain an equation, which is independent of the size of the electrode.

$$ i = i_o \left(\exp \{ \alpha_a F\eta / RT \} - \exp -\{ \alpha_c F\eta / RT \} \right) \tag{4} $$

In the Butler-Volmer equation the exchange current density, i_o, as the quantity describing charge exchange per unit surface at equilibrium potential in cathodic and anodic direction, is an indicator for electrocatalysis. Table 1 collects the well-established charge exchange current densities observed at different electrode materials whose numerical values range from 10^{-3} down to 5×10^{-13} Ampere per square centimetre (A cm^{-2}) over more than twelve orders of magnitude. Obviously the metals contained in the group of the platinum metals (Pd, Pt, Rh, Ir and, not reported here, also Ru) at which the hydrogen reaction is recorded to have the highest exchange current densities, do catalyse this reaction. They are good electrocatalysts for this reaction, whereas mercury on which an exceedingly low exchange current density is reported is not an electrocatalyst at all[6].

Table 1: Exchange current densities of the hydrogen evolution/anodic oxidation reaction at different electrode materials in aqueous 1 N H_2SO_4 solution at ambient temperature

Metal	io/A cm-2
Palladium, Pd	1.0 x 10-3
Platinum, Pt	8.0 x 10-4
Rhodium, Rh	2.5 x 10-4
Iridium, Ir	2.0 x 10-4
Nickel, Ni	7.0 x 10-6
Gold, Au	4.0 x 10-6
Tungsten, W	1.3 x 10-6
Niobium, Nb	1.5 x 10-7
Titanium, Ti	7.0 x 10-8
Cadmium, Cd	1.5 x 10-11
Manganese, Mn	1.3 x 10-11
Thallium, Ti	1.0 x 10-11
Lead, Pb	1.0 x 10-12
Mercury, Hg	0.5 x 10-13

Cyclic Voltammetry at Smooth Platinum, an Indicator of H-electrosorption and Electrocatalytic Activity for Hydrogen Evolution

The cyclic voltammogram of a platinum electrode immersed in acid electrolyte is depicted by broken line in Figure 4, which would be very similar in any electrolyte of any pH, provided that electrosorption of components of the electrolyte on platinum can safely be excluded[23]. One observes a relatively flat anodic and a very steep cathodic current increase at the right and left hand side of the plot, which are due to anodic oxygen and cathodic hydrogen evolution. The peaks and waves between electrochemical gas evolutions are due to electrochemical surface reactions at the platinum electrode. The voltammogram is observed by sweeping continuously the electrode potential with constant velocity of, for instance, 10 mV per second and thereby measuring the electrode current. Positive current peaks or waves do mean anodic oxidation processes and negative currents are correspondingly due to cathodic reactions at the platinum/electrolyte interface. A constant current between 0.3 and 0.6 V signify charging of the electrode/electrolyte interface, which forms a capacitance. The electrode potential is measured with reference to the Reversible Hydrogen Electrode (RHE) in the applied electrolyte. Beginning at 0 V vs. RHE, where the observed current is zero, one observes two oxidation peaks one centred around +150 and the other around +300 mV, which are attributed to the anodic oxidation of adsorbed hydrogen (H_{ad} $^{I, II}$ ® H^+ + e^-), which due to the respective different adsorption sites are adsorbed with different strengths. At 150 mV the free energy of adsorption corresponds to −14.5 kJ/mol and correspondingly the second peak signifies a free adsorption enthalpy of approximately −29 kJ per mol of adsorbed hydrogen atoms (reaction: $\frac{1}{2}H_2$ ® H_{ad}). These weak free adsorption enthalpies evidence that the binding energy of two H-atoms on platinum almost matches the dissociation/binding energy of the H-H bond. Neglecting for the moment the further anodic details of the voltammogram at more anodic potentials one observes on reversal of the voltage sweep cathodic currents and electrosorption peaks of hydrogen (H^+ + e^- ® H_{ad}), which mirror the discussed electrodesorption peaks at +300 and +150 mV. This evidences that the electrochemical reaction H^+ + e^- ® H_{ad} is a fast and reversible reaction on the time scale of the voltage sweep. These cathodic electrosorption and anodic electrodesorption peaks signify the pronounced adsorptive properties of platinum. Similar peaks are also observed at electrodes of other platinum metals, which are known to adsorb hydrogen[23].

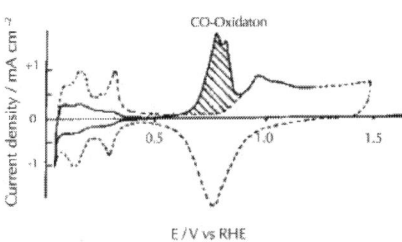

Figure 4: Cyclic voltammograms of a platinum electrode in acid solution (broken line) and CO-poisoned demonstrating anodic stripping of CO above

0.6V vs. RHE due to oxidation of adsorbed CO by anodically formed Pt-oxide.

Proceeding further with the sweep of the electrode potential to more positive potentials one passes through a region, where only a weak and constant anodic current is observed which signify only charging the capacity of the electrode double layer. Then a little bit above 600 mV the anodic current begins to rise approaching a current niveau above 800 mV. This increasing anodic current above 600 mV is connected to the beginning of surface oxidation of platinum (Pt + H_2O ® PtOH $_{surface}$ + H^+ + e^-), which forms a closed Pt oxide layer at approximately +800 mV. The almost constant anodic current, which is observed further, at still higher potential, serves then only to increase the thickness of the oxide layer and to increase the degree of oxidation of Pt (increasing the formal valency of the metal from +1 to higher values). Above 1.4 V, well above 1.2 V, the equilibrium potential of the oxygen electrode – the anodic current rises again and exponentially with electrode potential due to the onset of anodic oxygen evolution which shows a remarkable anodic overpotential because the reaction H_2O ® ½ O_2 + $2H^+$ + $2e^-$ is slow, kinetically hindered and Pt – oxides, which cover the electrode surface, are not particularly active catalysts to enhance it. Reversing the sweep leads to fast decline of anodic oxygen evolution currents and surface oxide charging currents. A broad cathodic current peak shifted negatively with respect to 0.6 V, the onset potential of Pt oxidation demonstrates the retarded reduction of the Pt-oxides. They can only be reduced with some cathodic overpotential because these oxides are (due to aging and further oxidation at higher potential) no longer the same and more stable and chemically more inert than those, which had been formed initially between +600 and +800 mV. Then approaching closer to the hydrogen equilibrium potential (0 V vs. RHE) one observes the cathodic electrosorption peaks for cathodic deposition of adsorbed hydrogen in two different adsorption states, which had been discussed before. Almost exactly at 0 V the cathodic hydrogen evolution current commences to increase steeply and exponentially with increasing overpotential. This increase is much steeper than that of the anodic oxygen evolution. Electrochemical ad- and desorption and electrochemical hydrogen evolution can be slowed appreciably or completely suppressed by strongly adsorbed substances – compare for CO-adsorption in Figure 4. Three items are immediately extracted from this cyclic voltammogram:

- Hydrogen is relatively strongly adsorbed on platinum and this voltammogram reveals different adsorption states with different adsorption enthalpies – both with a fixed stoichiometry. In contact with aqueous electrolyte at least two different adsorption sites on polycrystalline Pt can be identified with a free adsorption enthalpy of roughly –15 and –30 kJ per mol of adsorbed H. This adsorption is a stoichiometric reaction with Pt: H = 1:1, which allows coulometric determination (Integral [i dt]) of the free surface of Pt of disperded platinum.

- At platinum hydrogen evolution, and electroadsorption/ electrodesorption are fast, strongly catalysed approximately reversible reactions demanding little, almost negligible overpotential. The assumption of microscopic reversibility for this reaction on Pt/electrolyte interfaces is justified, provided this interface is clean and not blocked by electrosorbed species of any kind.

- In contrast, the anodic oxygen evolution does not proceed at the metallic Pt/electrolyte interface because this Pt metal is almost completely covered by Pt oxides already at +800 mV vs. RHE that means already at a potential more negative than 400 mV compared to the oxygen equilibrium potential. On these Pt oxides, the oxygen evolution and reduction is only weakly catalysed and comparatively much slower than cathodic hydrogen evolution on metallic platinum. Since the double bonded oxygen molecule is relatively stable, neither its reductive splitting nor its anodic formation is fast, must be particularly catalysed and cannot be expected to be a microscopically reversible process.

Volcano Plots Correlating Me-H Bond Strengths and Catalytic Activities

An approach to explain the particular role of platinum as an electrocatalyst for anodic hydrogen oxidation and cathodic hydrogen evolution is correlating the catalytic activity of different metals for this reaction and the strength of the adsorption enthalpy of hydrogen on these metals in Figure 5 [6]. In this figure the current densities measured for cathodic release of hydrogen at a given overpotential are plotted vs. the strength of the Me-H bond (calculated in kJ/mol). The reader may be reminded, that free H–adsorption enthalpies for immersed electrodes, as determined from a voltammogram and Me-H bond strength are not identical but related to each other. Figure 5 reveals in the so-called volcano curve a maximum of obtained current densities and hence electrocatalytic activities for intermediate Me-H bond strengths. That means that with lower bond strengths the interaction between adsorbate and metal is too low to allow for an effective activation of the H-H bond whereas with high Me–H bond strengths the adsorbed hydrogen is bonded too strongly to the electrode surface, so that is inactivated with respect to further reaction – be it further anodic oxidation in a fuel cell or reaction to liberated H_2 in cathodic hydrogen evolution. Therefore the Me-H binding energy of two hydrogen atoms approximately matches the binding energy of the H_2-molecule at most of the platinum metal group. It must be underlined that the volcano curve of Figure 5 is by far not restricted to electrochemical hydrogen reactions but is also observed for chemical reactions, for instance, for hydrogenation reactions of different unsaturated organic compounds and therefore is an indication for chemical catalysis of hydrogen splitting rather than for the charge transfer that is the electrochemical steps.

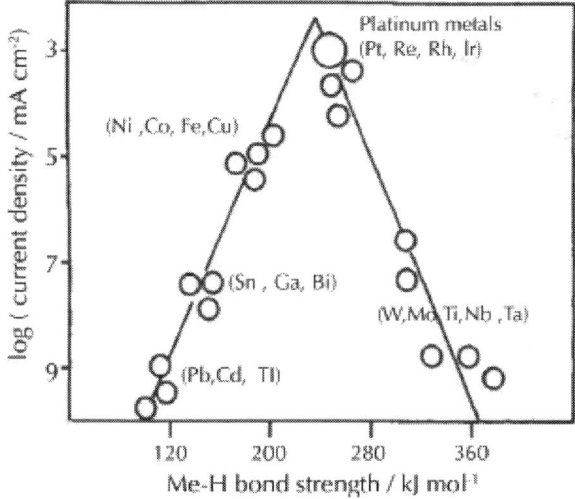

Figure 5: Volcano-type correlation of catalytic activity of different metals and the Me-H binding energy.

A different more fundamental approach to electrocatalysis of the anodic hydrogen oxidation consists in modelling hydrogen adsorption by theoretical chemical means, i.e., by calculations based on the functional density theory[24]. This approach aims at estimating binding energies of adsorbed hydrogen by calculating the surface specific number of empty d-electron states or d-orbitals respectively protruding from the metal surface towards the electrolytes (or for gas phase reactions towards the vacuum) at the metal surface. By taking into account the electronic binding energy of these d-orbitals of the metal (Fermi niveau) and the binding energy of electrons, coming from adsorbed molecules the energy of adsorption is calculated. This constitutes the strength of adsorption bonds of different adsorbates and allows calculating the binding energy of different adsorbed species, for instance hydrogen or carbon monoxide. For different platinum metal group and their alloys the results showed that hydrogen and carbon monoxide adsorption follow the same dependence that means, the stronger hydrogen is adsorbed, the stronger is also the adsorption of CO[25].

Electrocatalysis and Cocatalysts for CO-oxidation

Carbon monoxide, CO, is strongly adsorbed on platinum and other metals of the platinum group and it is particularly inert against anodic oxidation, although thermodynamically its oxidation potential is calculated at almost the equilibrium potential of hydrogen. Figure 4 shows a typical voltammogram of platinum electrode immersed in aqueous acidic solution that had been exposed to CO and was recorded after thorough purging of the electrolyte with Ar or N_2 or another inert gas, which expels CO from the electrolyte completely. Even after removal of the carbon monoxide from the solution, the platinum surface remains completely blocked by carbon monoxide[26]. This is first of all documented by the absence of the anodic peaks for adsorbed hydrogen, because of complete displacement of adsorbed hydrogen by much more strongly adsorbed carbon monoxide. Additionally one observes at anodic potential of approximately 0.6 V a relatively broad peak (typical for strongly hindered and slow electrochemical processes), which is due to anodic oxidation of the adsorbed carbon monoxide

(CO_{ad} + H_2O ® CO_2 + 2 H⁺ + 2e⁻). This process is called anodic stripping. And the anodic CO-stripping potential is kinetically – not thermodynamically – controlled proceeding with a high anodic overpotential. It is known, that this retarded oxidation is not a simple electrochemical reaction but that CO-oxidation is mediated by Pt-oxides, which are beginning to form at +0, 6 V vs. RHE. Therefore the observed anodic CO-stripping can be formulated in two steps (Equations 5a and 5b):

$$Pt + H_2O \rightarrow PtOH + H^+ + e^-$$

(5a)

$$CO_{ad} + 2\,PtOH \rightarrow CO_2 + 2\,Pt + H_2O$$

(5b)

The platinum oxide is immediately recuperated by anodic oxidation of the platinum surface at the anodic stripping potential. Only at such high overpotential the platinum surface is cleared of adsorbed CO and only then it becomes available for electrocatalyzed hydrogen oxidation. This means that CO-contaminated hydrogen can only be oxidized at anodic overpotential close to and above 0.6 V. A similar situation does also prevail with anodic oxidation of methanol, ethanol or other energy alcohols, because by their anodic conversion always strongly adsorbed CO is generated which poisons the platinum catalyst after very short time. Therefore metal-additives as cocatalysts are necessary which are more easily oxidized then platinum and whose oxides are able to oxidise adsorbed CO at much lower potential than would platinum oxide.

In Figure 6 is shown the current-voltage curves using an active carbon supported platinum anode for oxidation of pure hydrogen and hydrogen with 150 ppm of CO but using a modified platinum catalyst which contains other metals as a cocatalyst[27]. Clearly the presence of 150 ppm of CO almost destroys the catalytic activity of platinum for anodic hydrogen oxidation due to adsorptive poisoning as almost no current can be drawn from the anode. The addition of other metals, which by themselves are no H_2-oxidizing catalyst, as a potent cocatalyst in a Pt: Me (atomic ratio of 1:1) restores the catalytic activity of platinum. The effect is due to oxygen-transfer or spillover from metal oxides, which are already formed at close to +0.2 V. Interesting enough recent investigations of Baltrushat and co-workers[28] had shown that this effect is produced even if relatively small amounts of molybdenum are adsorbed on platinum.

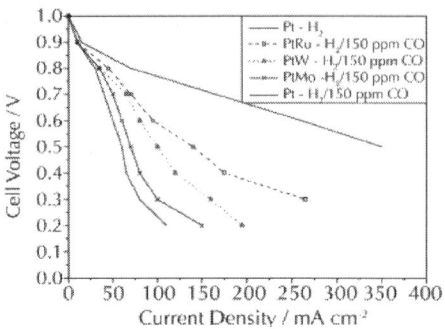

Figure 6: Current-cell voltage correlation of PEMFC with pure and COcontaminated hydrogen for supported pure Pt-electrocatalyst and binaryelectrocatalyst (Pt-loading 0.4 mg cm-2, 75 oC, 1 bar).

The Bifunctional Electrocatalyst Formulation

The first and practically still most important binary catalyst formulation for anodic reformate and methanol conversion is ruthenium which had already been mentioned in the famous conference report edited by Sandstede in 1968[29]. The remarkable cocatalytic efficiency of ruthenium is based on the observation, that this metal is less noble than platinum and oxidizes at a potential at least 0.2 V more negative than platinum and that ruthenium surface oxides/hydroxides therefore can act as a chemical oxidant for CO adsorbed on Pt at a potential, which is between 0.3 and 0.4 V more negative than anodic stripping from pure platinum would demand – provided ruthenium surface oxides and platinum are directly neighboured as would be the case in a surface alloy or if at least Pt and Ru particles would touch each other and the surface diffusivity of CO on platinum is high enough[30], which is the case. Therefore in this case the composite catalyst is called to be "bifunctional", as the platinum due to its adsorptive properties cares for bond breaking in H-H, O-H and C-H groups, whereas ruthenium in the form of its surface oxides mediates the chemical oxidation of the strongly adsorbed poison CO. But ruthenium is not singular and quite a number of elements particularly transition elements like Mo, W and also Ni, all being much less noble than Pt can act as cocatalysts in exactly that sense. The total mechanism is very similar to the anodic stripping reaction of Equation 5, as given by Equations 6a and 6b,

$$Me + H_2O \rightarrow MeOH + H^+ + e^- \tag{6a}$$

$$CO_{adsorbed\ on\ Pt} + 2\ MeOH \rightarrow CO_2 + Me + H_2O \tag{6b}$$

where, Me could be any metal being adsorbed on or alloyed with platinum and which is oxidized at potentials between 0 and 0.6 V vs. RHE as, for instance, Mo, W or others. Table 2 collects some of the alloying elements that had been shown to possess cocatalytic properties in binary Pt-metal electrocatalyst formulations[31]. A particular case is tin. Tin forms a well defined alloy Pt_3Sn, which is an active and particularly efficient cocatalyst for CO-oxidation[32], whose use in membrane fuel cells might become state of the art within the next years.

Table 2: Composition of various CO-tolerant binary Pt-based electrocatalysts

Adittive	Composition
Ru	Pt35Ru65
	Pt57Ru43
Co	Pt57Co43
	Pt82Co18
Ni	Pt58Ni52
	Pt76Ni24
Fe	Pt85Fe15
	Pt95Fe0,5
Sn	Pt3Sn
Mo	Pt45Mo55
	Pt67Mo33

Anodic Oxygen Evolution in Electrolysis Processes and Cathodic Oxygen Reduction in Fuel Cells

Anodic oxygen evolution in aqueous solutions applied, for instance, in water electrolysis or metal electrowinning processes, Equation 7,

$$H_2O \rightarrow 2H^+ + \tfrac{1}{2}\,O_2 + 2\,e^-$$

$$(7)$$

proceeds always at metal oxide covered metal electrodes, the anodic overpotential usually exceeding 0.2 V and is supposed to proceed according to the so-called Krasilch'shikov mechanism[6], in which unstable, overoxidized metal oxide sites are self stabilizing by mutual redox- or disproportionation reactions by release of molecular oxygen (schematically described by Equation 7a and 7b) with regeneration of the lower valent metal oxide:

$$Me_xO_y + H_2O \rightarrow Me_xO_yOH + H^+ + e^-$$

$$(7a)$$

$$4\,Me_xO_yOH \rightarrow 4\,Me_xO_y + O_2 + 2\,H_2O$$

$$(7b)$$

This mechanistic interpretation is based on the observation, that searching for a volcano-like correlation the correct quantity for correlating different catalytic activities has little to do with adsorption of oxygen. It is simply the free enthalpy of formation of the overoxidized metal oxide sites – namely the free enthalpy of formation of the higher oxide from the stable metal oxide. The maximum, the tip of the volcano, is observed, where the equilibrium potential for the lower valent and higher valent metal oxides (which can be calculated from the Gibbs-enthalpy of the oxidation reaction of the metal oxide) matches that of the equilibrium potential of the oxygen electrode (+1.2 V vs. RHE)[33].

Mechanism of the Cathodic Oxygen Reduction

Because of non-existing microscopic reversibility the cathodic reduction of oxygen follows, however, quite a different route than oxygen evolution. By adsorptive interaction of O_2 with the noble platinum metals the molecule is activated towards bond-splitting and electron uptake. The best electrocatalyst for cathodic oxygen reduction hitherto known is platinum metal. Referring to the cyclic voltammogram of Figure 4, one observes the formation of surface oxides on a platinum electrode above +600 mV and the anodic current reaches a plateau above +800 mV as an indication of the formation of an almost closed platinum oxide layer that grows steadily. But even at +800 mV still a minute fraction of the metal surface is not covered by Pt oxide (PtOH) but it becomes less and less with increasing the electrode potential. The degree of coverage, Q, in a simplified correlation, follows a type of electrochemical Langmuir isotherm, Equation 8a and 8b

$$\Theta / (1- \Theta) = K_{ox} \tag{8a}$$

$$K_{ox} = \exp (E - E_{1/2}) F / RT \tag{8b}$$

where, Q is the degree of coverage of the surface by the oxide and K_{ox} is the potential-dependent equilibrium constant. $E_{1/2}$ is the potential at which half of the surface is covered by the oxide. Practically this is the potential at half of the surface oxidation current is measured (approximately 0.7 V). Equation 8a and 8b are only to be understood in a schematic way, but they demonstrate the most important fact, that increasing the anode potential still leaves a smaller and smaller but still finite rest of the platinum surface uncovered by oxide. Therefore oxygen adsorption and reduction is not a question of yes or no but of the degree of Pt-oxide coverage, which depends for a chemical equilibrium reaction on the applied electrode potential.

The adsorptive interaction of electron-donating double bonded oxygen and empty d-orbitals at the Pt-surface acting as acceptors with the ability of d_p - p_p^* back donation activates the oxygen molecule to an extent which allows for a fast transfer of four electrons, a fast sequence of several electron transfer and chemical steps, which are today not yet completely elucidated. This explains, why a relatively high cathodic overpotential of at least $- 0.2$ V is necessary in order to start with cathodic oxygen reduction at Pt–electrodes, as a rough calculation on the basis of Equation 8a clarifies, that the fraction of free metallic surface atoms amounts to roughly 10^{-3} at 1 V vs. RHE ($- 0.2$ V vs. oxygen equilibrium).

Electrocatalyzed oxygen reduction proceeds at Pt-electrodes only on metal–sites in the adsorbed state and two parallel reaction pathways are observed in acid solution: the direct reduction of O_2 to H_2O (Equation 9a) and an indirect pathway with H_2O_2 (Equation 9b and 9c)[26, 34]:

Direct reduction

$$O_2 + 4H^+ + 4e^- \rightarrow 2H_2O \; E^0 = 1.23 \text{ V} \tag{9a}$$

Indirect reduction

$$O_2 + 2H^+ + 2e^- \rightarrow H_2O_2 \; E^0 = 0.682 \text{ V} \tag{9b}$$

$$H_2O_2 + 2H^+ + 2e^- \rightarrow 2H_2O \; E^0 = 1.77 \text{ V} \tag{9c}$$

In order to the oxygen reduction reaction to take place, the oxygen bond must be weakened, which in turn implies that a strong interaction with the surface of the catalyst will be necessary. If the adsorption of the O_2 molecule is end-on, this does not weaken the O-O bond sufficiently, and the indirect mechanism is favoured, whereas bidentate bonding of the O_2 to a surface metal atom, or bridge bonding between two surface metal atoms will favour the direct process, since in this way the O-O bond is substantially weakened[26,34]. The 2-electron reduction of oxygen to hydroperoxide with still intact O-O bond ($O_2 + 2e^- + H_2O$ ® H_2O_2) is produced, for instance, by O_2–reduction to hydroperoxide (H_2O_2) on graphite electrodes in an industrial process in the pulp and paper industry as bleaching agent.

HIGHLY DISPERSE PLATINUM AND PLATINUM ALLOYS: THE ELECTROCATALYST FOR PHOSPHORIC ACID AND MEMBRANE FUEL CELLS

Dispersed Platinum Invented for Phosphoric Acid Fuel Cells

Already in the late 70ies of the last century UTC (United Technology Corporation) developed highly disperse platinum supported on active carbon as anodic (H_2 oxidation) and cathodic (O_2 reduction) electrocatalyst for phosphoric acid fuel cells[10]. Much older had been the use of so-called platinum black as a form of – although not very well defined – disperse platinum metal for catalytic purposes for instance in the Döbereiner lighter (the oldest and very popular application) and for catalytic hydrogenations in synthetic organic chemistry. Already in this very early times of the 18[th]century existed the idea to increase the effective catalytic activity of the metal catalyst by dispersing the material as finely as possible. The relevant idea was to make highest possible use of the precious metal by dividing it into very small crystallites. Thus the relevant surface to volume ratio according to Equation 10 for a globular particle can be approximated by 3/r, with r being the particle radius. Thus for a particle of 1 nm (10^{-9} m or 10 Å) the surface to volume ratio amounts to 10^7 cm^2/cm^3 = 10^7 cm^{-1}. Even more impressive than this number is the knowledge, that in a globular platinum particle of approximately 2 nm diameter roughly 1/3 of all platinum atoms are exposed to the surface and only 2/3 are buried in the interior of this particle.

$$(\text{Surface/Volume})_{\text{spere}} = 4\Pi r^2 / 4/3\ \Pi r^3 = 3/r \qquad (10)$$

Practically the utilisation of platinum cannot be expected to be much greater in a particle intended to be stable long term in an operating fuel cell. Highly dispersed materials are thermodynamically unstable. Due to their high surface energy they tend to agglomerate, which would decrease the excess surface energy. Hence fast deactivation of the catalyst due to "sintering" would occur. Anchoring such particles by adsorption on the surface of active carbon allows separating these particles from each other stabilizing them long-term. Figure 7 shows a micrograph of such active carbon-supported platinum-ruthenium nanoparticles obtained by transmission electron microscopy[35]. The carbon supported Pt catalysts prevents the fast agglomeration of the nanoparticles, a process that would occur with platinum black, because the platinum particles are touching each other.

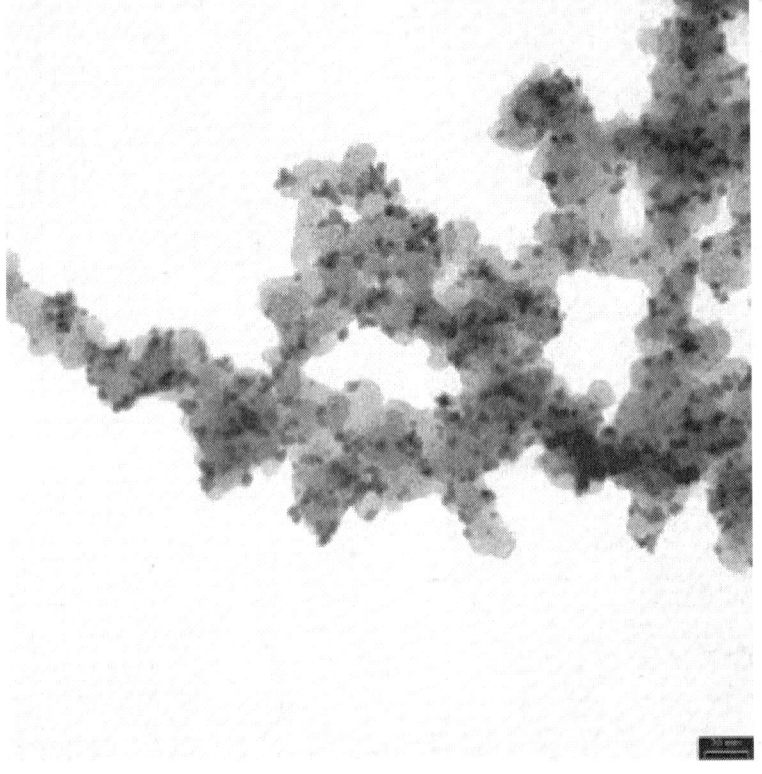

Figure 7: Transmission electron micrograph of PtRu/C electrocatalyst (20wt%, Pt: Ru molar ratio 1:1) prepared by alcohol-reduction process.

INDUSTRIAL ACTIVE CARBON SUPPORTED ELECTROCATALYSTS, THEIR PROPERTIES AND PREPARATION

Properties of Active Carbon

Active carbon is a technical mass product, a type of soot, obtained by incomplete combustion and gas phase pyrolysis of hydrocarbons, unsaturated hydrocarbons as ethene (ethylene) and ethyne (acetylene) and aromatic compounds in particular[11]. The production procedure secures reproducible qualities both with respect to specific surface and with respect to surface chemistry. Active carbons come in different qualities and most frequently are amorphous and possess an inner surface of the order of from several tens to several 100 m^2 per gram. Two of the most popular qualities are Black Pearls 2000® with approximately 1.300 m^2 per gram and Vulcan® XC-72 with 250 m^2 per gram both from Cabot Corp. Literally there are hundreds of different active carbon qualities available on the market. The surface of active carbons is to some extent oxidized, carrying hydroxyl-, carbonyl-, carboxylate-, ester-, ether- and quinone-groups and unsaturated groups _ sometimes also nitrogen. Upon oxidation by strong oxidants like chromic acid they obtain cation exchange properties[11]. Active carbons are, due to their high inner surface, thermodynamically unstable with respect to bulk graphite but non the less they are so inert, that they resist reductive and oxidative attack and form a material, which is long term stable in low temperature fuel cells even at the cathode side, where the working potential is oxidative (+800 mV vs. RHE). Their practical stability is accounting for life times of 20.000 h (vehicles) or even 40.000 h at 200 °C in phosphoric acid fuel cells. It should be kept in mind, that for instance in phosphoric acid fuel cells on the cathode, half of the supporting active carbon disappears by anodic oxidation within 40.000 h of service time or approximately within four years.

Preparation of Dispersed Platinum and Active Carbon Supported Platinum

The preparation of highly dispersed metals proceeds according to two principally different lines[36,37]:

- On one hand the metal can be prepared separately as a sol or stabilized, solvated (not precipitated) nanodisperse material by deposition of the small metal particles from aqueous solutions of respective salts. Most commonly the precipitation is accomplished by strong reductants and assisted by the presence of surface-active substances, which help to prevent growing of the crystals beyond a certain size.

- On the other hand the dispersed metal particles can be obtained on the active carbon surface after impregnating the material with noble metal salt solutions, slow drying with salt deposition on the active carbon followed by reduction. The reductive treatment may be most commonly performed by dissolved reductants as formaldehyde or hydrazine or may be performed by flowing hydrogen or other gaseous reductants at relatively low temperature (200-300 °C). As a reductant one may also use the supporting active carbon proper at higher temperatures exceeding 400 °C. By this way the so-called cabothermic reduction of the metal salts on active carbon can be performed.

Concerning these two lines, the preparation of highly dispersed platinum have been done by different approaches[37]:

- Sols of the metal are obtained for instance by an initial formation of a metastable platinum – sulfito complex, which is inert at ambient temperature but decomposes and produces the small Pt-crystallites at temperatures in excess of 60 °C. Thus a relatively well-defined crystal size of from 2 to 6 nm can be obtained. Bringing then active carbon into contact with the sol «solution» results in anchoring of the metal crystallites on the carbon surface.

- Impregnating the active carbon by aqueous or nonaqueous, organic solvents containing platinum salts assisted by the addition of dissolved wetting agents, followed by drying the impregnated material leads to a relatively homogeneous distribution of platinum salts on the active carbon material. Successively reducing these salts, for instance with hydrogen at relatively low temperatures allows creating nanoscale platinum particles whose diameter vary, depending on reduction and preparation conditions, between 2 and 8 nm.

- Chemical oxidation of active carbon by strong oxidants as for instance by aqueous solution of chromic acid creates acidic groups (OH-groups, phenolic, and carboxylate groups) on the active carbon surface, which make the active carbon a cation exchange material. From aqueous solutions, which contain positively charged platinum complexes and entities (for instance, amino-complexes of Pt^{+2}), these moieties are absorbed and bound to the active carbon by ion exchange. Chemical or (less often used) electrochemical reduction produces the dispersed metal on active carbon.

Platinum Alloys and Composite Multi-metal Catalysts for the Anodic Oxidation of Carbon Monoxide Containing Reformate, of Methanol and Other Energy Alcohols

As documented by a considerable number of patents already in the late 70ies the developers of phosphoric acid fuel cells had detected that from highly dispersed platinum on active carbon, dispersed platinum alloys can be obtained by impregnating of the catalyst with aqueous solutions of transition metal salts and subsequent drying and heat treatment[36]. Subsequent leaching of the active carbon with hydrochloric acid removes excesses of non-alloyed metal oxide and leaves on the support the platinum-transition metal alloy. At that time alloy formation, which is still in habit for PAFC electrocatalysts had the main purpose of catalyst stabilisation. It is known today, that on Pt-transition metal alloys also adsorption of oxygen and its catalytic activation is enhanced compared to pure platinum. Therefore the alloy catalyst is not only more stable but also additionally more active because due to alloying the electronic structure of platinum is changed in a way that favours O_2-adsorption on platinum. In the context of reformate and alcohol consuming fuel cells alloy formation and catalyst formulations of binary and ternary composition became indispensable because cocatalysts are needed to oxidize carbon monoxide and other, strongly adsorbed oxidation- and oxidative degradation products of methanol or the higher alcohols together with hydrogen[38,39]. Ralph and Hogart[40], two scientists of Johnson Matthey published the history of particle stabilisation by alloying (for the cathode) and CO–resistance enhancement by alloying (for the anode) till 2001 and Thompson[41] published a respective article in the Handbook of Fuel Cells.

The Bönnemann Method[42]

Preparing nanodisperse electrocatalysts which are composed of two, three or even more elements, demands that the preparation method produces not a mixture of crystals of the different elements but a well-defined molecular mixture – if possible well defined alloys of the elements in the electrocatalyst particles of nanometer size. By precipitation from aqueous solutions this cannot be attained. Therefore, recently the procedure of Bönnemann became very popular. It is essentially based on reductive precipitation of metals from their salt solutions in aprotic media by a borohydride, a very strong reductant. The intention is the forced codeposition of metals irrespective of their redox potential or different nobleness respectively. The tetrabutyl ammonium cation is a strong surfactant that adsorbs on the precipitated crystallites and prevents their growing beyond a size of between 2 and 3 nm. The result is a nanodisperse material in the form of a sol, which then is anchored by adsorption on active carbon as the usual support. Effectively the reductant is tetrabutyl-tridodecyl-borohydride $N(but)_4$ $HB(C_{12}H_{25})_3$. Precipitation for instance of a 1:1:1 mixture of Pt, Ru and W is performed in tetrahydrofuran solution containing dissolved H_2PtCl_6, $RuCl_3$ and WCl_3 in equimolar amounts[43]. By this method almost any desired mixture of elements in any desired molecular ratios can be produced. However, recent investigations show, that not always the desired mixture on molecular level can be obtained as the different reducibility of compounds of different elements occurs sometimes with different rate, so that in the precipitate a certain segregation of the elements is non-voluntarily obtained. Therefore it is clear, that mixed electrocatalysts produced by the Bönnemann method must always be investigated, whether alloys or segregated nanoparticles of variable compositions have been obtained and what is their microstructure and elemental composition.

In Brazil, the electrocatalysts produced by the Bönnemann method with binary, ternary and quaternary compositions have been studied at IPEN/CNEN-SP[44-50](Instituto de Pesquisas Energéticas e Nucleares). Other methodologies of electrocatalysts preparation have also been studied and developed at IPEN/CNEN-SP[51-55] and IQSC/USP[56-61](Instituto de Química de São Carlos, Universidade de São Paulo).

Means and Methods Characterizing Dispersed Platinum and Platinum Alloy Catalysts

Cyclovoltammetry as a method to characterize the electrochemical properties and electrocatalytic potential of active carbon supported nanodisperse electrocatalyst can be applied in a similar manner as is usual for solid electrodes. It is possible to fix the catalyst on glassy carbon by a submicrometer layer of Nafion[62]. This, however, allows only a measurement on short term for screening the most potent formulations among a preselection of a number of different but similar catalyst formulations. The ultimate electrochemical characterization is the long-term performance test in real fuel cells under typical operation conditions. Here long-term means at least several hundred hours, if not more than 1000 h. For research and development, however, many questions concerning the nanostructure of the electrocatalyst must be asked and answered as:

• Size and size distribution of the catalysts.

• Are the components in binary and ternary catalyst formulations attributed to Pt in a form of alloy?

• Are the additives evenly distributed on the support and in what chemical form and environment and what is their valency?

• If not alloyed, are the additives present in well-defined crystalline phases or are they X-ray amorphous? And how are they spatially distributes.

The most important method to answer these questions is by Transmission Electron Microscopy (TEM). This method fails with X-ray amorphous material but with well-crystallized catalysts it allows for size and size distribution and even determination of lattice parameters. X-ray diffraction according to Debye-Scherrer answers almost the same questions as widening of the refraction signals gives information about the size of the refracting particles. Another method allowing gross analytic identification of the elements and their relative amounts in the catalyst is XPS (X-ray Photoelectron Spectroscopy) and EDX (Energy Dispersive X-ray Spectroscopy). These methods are usually integrated in scanning electron microscopic equipment[63]. All these methods are typical ex-situ methods as they are performed in separate vacuum-apparatus. In-situ characterisation of the solid state properties of the electrocatalyst is only possible by X-ray absorption of "white" X-ray light which is only available in the form of synchrotron-radiation[64]. Although theory and practice of X-ray Absorption Spectroscopy (XAS) are complex and too much specialised to be treated in the context of this article it should be stressed that a great many questions in particular questions ii), iii) and iv) can be answered by the different forms of XAS. The second in-situ method is IR-Absorption Spectroscopy, which tells many things about specifically adsorbed species but almost nothing about the chemistry and physics of the adsorbent, the catalyst or parts of it.

FINAL CONSIDERATIONS

What will be the technological future of low temperature fuel cells and the respective Research and Development into electrocatalysts and electrocatalysis? For the time being, a certain deceleration of research and development is observed because the hopes for a fast evolving fuel cell automobile market could not be fulfilled. But the task of catalyst research, development and fabrication for low temperature fuel cells is far from being finished. The salient point with fuel cell cars was not technical feasibility but costs – which postponed fuel cell development for cars to the late twenties – and then it will be hydrogen/air not methanol reformate/air cells. Nevertheless with so-called portables running on direct methanol cells, APUs (auxiliary power sources for heavy duty vehicles, ships and planes) running on reformate from gasoline and domestic electricity and heat cogeneration based on natural gas there is ample demand for better and more stable anodic and cathodic electrocatalysts. Small portable fuel cells are operated on liquid methanol (more exactly approx. 1 mol L^{-1} aqueous solutions of methanol). Portables are small cells with electric power of 10 to 100 W, which supply camcorders, PCs and other electronic devices. Portables are in this specialised field with costs of from 15,000 to 18,000 USD/kW cost-effective enough to compete with high-valued batteries because their capacity is remarkably higher (charge lasts for approx. a day instead of 3 to 4 h). In the Internet there are some addresses as for instance Smart Fuel Cell® or Oksolar®, to name only two, which testify commercial activities and success.

Auxiliary Power Units (APU) are another interesting option for cars, ships and aeroplanes, which are operated on reformed gasoline or Diesel. They are supposed to substitute heavy and expensive lead acid batteries. For them the same or very similar catalyst formulations would be needed as for reformate powered cells. Finally domestic power sources, which provide electricity and heat for household fired with natural gas may become an option in the future.

For all these purposes PEM – cells have been and still are being developed. There are from the technical and fuel point of view two very different lines of development: PEMFCs for portables with aqueous methanol solutions as fuel which work at close to ambient temperature and PEMFCs cells for APUs, domestic and other stationary power and heat generation which work on reformate gas at enhanced (though low) temperature around 80 °C or somewhat higher. The demands with respect to the anodic electrocatalyst are very different for these two different technologies. Using as fuel aqueous methanol solution, the stability with respect to temperature needs not to be very high, but long-term stability together with stability against frequent current interruption, fuel starvation and voltage-cycling is very important. Additionally the cathode poses a particular problem. With liquid methanol at the anode, substantial amounts of methanol are leaking to the cathode whose performance is severely impaired by methanol poisoning. There are today interesting approaches to make the cathode less sensitive towards methanol poisoning by applying Ru and Pt containing Chevrel phases – but still the state of the art is rudimentary in this respect and the same holds with the anode catalyst, where still nothing better than Pt/Ru is being applied.

PEM cells for domestic use or for APUs working on reformate can still be improved by more sophisticated anodic electrocatalysts as discussed in this article. It is therefore expected, that intensive R&D research will continue on the still relatively young field of fuel cell electrocatalysis and that many innovations and advances are still ahead. Also production technologies must be developed for new catalysts.

Very important is it to take into account, what will be the advances in membrane-technology in the near future. Intensive work worldwide aims in finding a substitute for the usual Nafion membrane and to introduce less solvated ionomers into fuel cell technology in order to raise the working temperature of the gas-operated cell substantially above 100 °C (even to 200 °C). Clearly, higher operating temperatures demands new electrocatalysts. We do now know a lot about activity enhancement by alloying and this knowledge can also be accounted for in theoretical chemistry – but this has not yet predictive power, so that still experimentation, flanked by the marvellous extended analytical methods will be the usual tools of research and development.

Last but not least it should be mentioned that ethanol is a special and typical fuel option for Brazil. Be it reformate from ethanol for PEM-cells or direct anodic oxidation of ethanol in PEM-cells, for both purposes highly active and selective electrocatalysts are indispensable[65]. In Brazil, the direct anodic oxidation of ethanol has been studied at IPEN/CNEN-SP[53-55] and IQSC-USP[66,67], using electrocatalysts prepared by different methodologies. Pt/Sn electrocatalyst has been shown a performance superior to that of Pt/Ru electrocatalyst for direct anodic oxidation of ethanol (Figure 8)[68].

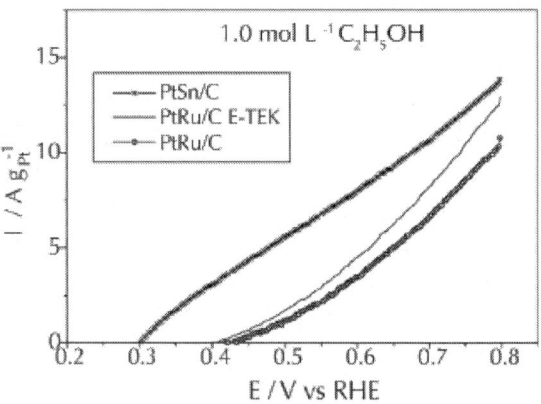

Figure 8: Cyclic voltammetry of PtRu/C and PtSn/C electrocatalysts (20wt%,Pt:Me molar ratio 1:1) prepared by alcohol-reduction process and the commercial PtRu/C E-TEK electrocatalyst (20wt%, Pt: Ru molar ratio 1:1) in 0.5 mol L^{-1} H$_2$SO$_4$ containing 1.0 mol L^{-1} of ethanol with a sweep rate of 10 mV s^{-1}, considering only the anodic sweep.

ACKNOWLEDGEMENTS

H. Wendt is indebted to IPEN/CNEN-SP, to the Brazilian Ministry of Science and Technology and to FINEP for obtaining a scholarship at São Paulo in the year 2003, when he spent four weeks lecturing on *Electrocatalysis* and on *Modern Energy Technologies* as guest lecturer with the group of M. Linardi. His and his collaborators kind hospitability is gratefully acknowledged.

Financial supports of CTPETRO/FINEP and FAPESP (Fundação de Amparo a Pesquisa do Estado de São Paulo) for the research activities at IPEN/CNEN-SP are greatly acknowledged.

REFERENCES

1. Carrette, L.; Friedrich, K. A.; Stimming, U.; *Fuel Cells* 2001, *1*, 5.

2. Stull, D. R.; Prophet, H.; *Janaf Thermochemical Tables*, 2nd ed., US National Bureau of Standards: Washington, 1971.

3. Barin, I.; Knacke, O.; *Thermodynamical Properties of Inorganic Substances*, Springer Verlag: Berlin, 1973

4. Kauffmann, G. B.; *Platinum Metals Rev.* 1999, *43*, 122.

5. Wendt, H.; Rausch, S.; Borucinski, T.; *Adv. Catal.* 1994, *40*, 87.

6. Wendt, H.; Kreysa, G.; *Electrochemical Engineering, Science and Technology in Chemical and Other Industries,* Springer: Berlin, 1999, ch. 4 and 10.

7. Kemp, F.S.; George, M. A.; *US pat. 3,857,737* 1974.

8. Petrov, H. G.; Allen, R.J.; *US pat. 3,992,512* 1976.

9. Kinoshita, K.; Stonehart, P. In *Modern Aspects of Electrochemistry*; Bockris, J. O'M.; Conway, B. E., eds.; Plenum Press: New York, 1977, vol. 12.

10. Stonehart, P.; *Ber. Bunsen Phys. Chem.* 1990, *94*, 913.

11. Kinoshita, K.; *Carbon – Electrochemical and Physical Properties*, Wiley and Sons: New York, 1987.

12. Bacon, F. T.; *Electrochim. Acta* 1969, *14*, 569.

13. Ketelaar, J. A. A. In *Fuel Cell Systems;* Blomen, L. M. M. J.; Mugerwa, M. N., eds.; Plenum Press: New York, 1993.

14. Winsel, A.; Richter, G. J. In *Alkaline Fuel Cells in Electrochemical Hydrogen Technologies*; Wendt, H., ed.; Elsevier: Amsterdam, 1990

15. Watkins, D. S. In ref. 13.

16. Strasser, K.; *J. Electrochem. Soc.* 1980, *127*, 2172.

17. Steele, B. C. H.; Heinzel, A.; *Nature* 2001, *414*, 345.

18. Neergat, M.; Friedrich, K. A.; Stimming, U. In *Handbook of Fuel Cells – Fundamentals, Technology, Applications*; Vielstich, W.; Lamm, A.; Gasteiger, H. A., eds.; J. Wiley & Sons: New York, 2003, vol. 4, chap. 63

19. Acres, G. J. K.; Frost, J. C.; Hards, G. A.; Potter, R. J.; Ralph, T. R.; Thompsett, D.; Burstein, G. T.; Hutchings, G. J.; *Catal. Today* 1997, *38*, 393.

20. Roth, C.; Goetz, M.; Fuess, H.; *J. Appl. Electrochem.* 2001, *31*, 793.

21. Trasatti, S.; *Electrochim. Acta* 2003, *48*, 3729.

22. Debe, M. K. In ref. 18, vol. 3, chap. 45.

23. Vielstich, W. In ref. 18, vol. 2, chap. 14.

24. Liu, P.; Norskøv, J. K.; *Fuel Cells* 2001, *1*, 192.

25. Liu, P.; Logadottir, A.; Norskøv, J. K.; *Electrochim. Acta* 2003, *48*, 3731.

26. Hamann, C. H.; Hammett, A.; Vielstich, W.; *Electrochemistry*, Wiley-VCH: Weinheim, 1998.

27. Goetz, M.; Wendt, H.; *Electrochim. Acta* 1998, *43*, 3637.

28. Samjeske, G.; Wang, H.; Loftler, T.; Baltruschat, H.; *Electrochim. Acta* 2002, *47*, 3681.

29. Binder, H.; Köhling, G.; Sandstede, G.; *From Electrocatalysis to Fuel Cells*, University of Washington Press: Seattle, 1972.

30. Friedrich, K. A.; Geyzers, K-P.; Linke, U.; Stimming, U.; Stumper, J.; *J. Electroanal. Chem.* 1996, *402*, 123.

31. Watanabe, M. In ref. 18, vol. 2, chap. 28.

32. Gasteiger, H. A.; Marcovic, N. M.; Ross, P. N.; *J. Phys. Chem.* 1995, *99*, 8945.

33. Trasatti, S.; *Electrochim. Acta* 1984, *29*, 1503.

34. Tanata, A.; Kanamura, K.; Adzic, R.; Cahan, B.; Yeager, E.; *Proc.- Electrochem. Soc.* 1990, *90-1*, 969.

35. Spinacé, E. V.; Neto, A. O.; Vasconcellos, T. R. R.; Linardi, M.; *Br. Patent INPI-RJ* PI0304121-2,2003.

36. Jalan, V. M.; *US pat. 4,202,934* 1980.

37. Mukerjee, S.; Srinivasan, S. In ref. 18, vol. 2, chap. 34.

38. Torresi, R. M.; Wasmus, S. In ref. 18, vol. 2, chap. 15.

39. Iwasita, T.; *Electrochim. Acta* 2002, *47*, 3663.

40. Ralph, T. R.; Hogarth, M. P.; *Platinum Metals Rev.* 2002, *46*, 117.

41. Thomsett, D. In ref. 18, vol. 3, chap. 37.

42. Bönnemann, H.; Brijoux, W.; Brinkmann, R.; Fretzen, R.; Joussen, T.; Köppler, R.; Korall, B.; Neiterle, P.; Richter, J.; *J. Mol. Catal.* 1994, *86*, 129.

43. Goetz, M.; Wendt, H.; *J. Appl. Electrochem.* 2001, *31*, 811.

44. Neto, A. O.; Franco, E. G.; Aricó, E. M.; Linardi, M., Gonzalez, E. R.; *International Workshop on Ceramic & Metal Interfaces*, Oviedo, Spain, 2002.

45. Franco, E. G.; Oliveira Neto, A.; Linardi, M.; Aricó, E.; *J. Braz. Chem. Soc.* 2002, *13*, 516.

46. Neto, A. O.; Franco, E. G.; Aricó, E. M.; Raimundo, C. P.; Linardi, M.; *XV Congresso da Sociedade Ibero-Americana de Eletroquímica*, Évora, Portugal, 2002.

47. Oliveira Neto, A.; Franco, E. G.; Aricó, E.; Linardi, M.; Gonzalez, E. R.; *J. Eur. Ceram. Soc.* 2003, *23*, 2987.

48. Neto, A. O.; Franco, E. G.; Aricó, E. M.; Spinacé, E. V.; Linardi, M.; *Fuel Cells Science and Technology,* Amsterdam: Netherlands, 2002.

49. Franco, E. G.; Aricó, E.; Linardi, M.; Roth, C.; Martz, N.; Fuess, H.; *Mater. Sci. Forum* 2003, *416*, 4.

50. Franco, E. G.; Neto, A. O.; Raimundo, C. P.; Aricó, E. M.; Linardi, M.; *53rd Annual Meeting of the International Society of Electrochemistry*, Dusseldorf, Germany, 2002.

51. Spinacé, E. V.; Oliveira Neto, A.; Franco, E. G.; Linardi, M.; Gonzalez, E. R.; *Quim. Nova* 2004, *27*, 648.

52. Neto, A. O.; Franco, E. G.; Spinacé, E. V.; Linardi, M.; *54th Annual Meeting of the International Society of Electrochemistry*, São Pedro, Brasil, 2003.

53. Spinacé, E. V.; Neto, A. O.; Linardi, M.; *J. Power Sources* 2003, *124*, 426.

54. Spinacé, E. V.; Neto, A. O.; Linardi, M.; *J. Power Sources* 2004, *129*, 121.

55. Spinacé, E. V.; Neto, A. O.; Vasconcelos, T. R. R.; Linardi, M.; *J. Power Sources* 2004, *137*, 17.

56. Castro Luna, A. M.; Camara, G. A.; Paganin, V. A.; Ticianelli, E. A.; Gonzalez, E. R.; *Electrochem. Commun.*2000, *2*, 222.

57. Lizcano-Valbuena, W. H.; Souza, A.; Paganin, V. A.; Leite, C. A. P.; Galembeck, F.; Gonzalez, E. R.; *Fuel Cells*2002, *2*, 159.

58. Colmati Jr., F.; Lizcano-Valbuena, W. H.; Camara, G. A.; Ticianelli, E. A.; Gonzalez, E. R.; *J. Braz. Chem. Soc.*2002, *13*, 474.

59. Camara, G. A.; Giz, M. J.; Paganin, V. A.; Ticianelli, E. A.; *J. Electroanal. Chem.* 2002, *537*, 21.

60. Lizcano-Valbuena, W. H.; Paganin, V. A.; Leite, C. A. P.; Galembeck, F.; Gonzalez, E. R.; *Electrochim. Acta* 2003, *48*, 3869.

61. Lizcano-Valbuena, W. H.; Azevedo, D. C.; Gonzalez, E. R.; *Electrochim. Acta* 2004, *49*, 1289.

62. Schmidt, T. J. In ref. 18, vol. 2, chap. 22.

63. Hoster, H. E.; Gasteiger, H. A. In ref. 18, vol. 2, chap. 18.

64. Adzic, R. R.; Wang, J. X.; Ocko, B. M.; McBreen, J. In ref. 18, vol. 2, chap. 20.

65. Lamy, C.; Belgsir, E. M.; J-M. Leger, J-M.; *J. Appl. Electrochem.* 2001, *31*, 799.

66. Oliveira Neto, A.; Giz, M. J.; Perez, J.; Ticianelli, E. A.; Gonzalez, E. R.; *J. Electrochem. Soc.* 2002, *149*, A272.

67. Souza, J. P. I.; Queiroz, S. L.; Bergamaski, K.; Gonzalez, E. R.; Nart, F. C.; *J. Phys. Chem. B* 2002, *106*, 9825.

68. Spinacé, E. V.; Oliveira Neto, A.; Vasconcelos, T. R. R.; Linardi, M.; *Abstracts of the VI Encontro Regional de Catálise, Americana, Brasil*, 2004.

Chapter 5

Experimental Study of Gas Explosions in Hydrogen Sulfide-Natural Gas-Air Mixtures

André Vagner Gaathaug[1], Dag Bjerketvedt[1], Knut Vaagsaether[1], and Sandra Hennie Nilsen[2]

[1]Telemark University College, Faculty of Technology, 3918 Porsgrunn, Norway

[2]Research, Development and Innovation, Section for Health, Safety and Water Management, Statoil ASA, 3905 Porsgrunn, Norway

ABSTRACT

An experimental study of turbulent combustion of hydrogen sulfide (H_2S) and natural gas was performed to provide reference data for verification of CFD codes and direct comparison. Hydrogen sulfide is present in most crude oil sources, and the explosion behaviour of pure H_2S and mixtures with natural gas is important to address. The explosion behaviour was studied in a four-meter-long square pipe. The first two meters of the pipe had obstacles while the rest was smooth. Pressure transducers were used to measure the combustion in the pipe. The pure H_2S gave slightly lower explosion pressure than pure natural gas for lean-to-stoichiometric mixtures. The rich H_2S gave higher pressure than natural gas. Mixtures of H_2S and natural gas were also studied and pressure spikes were observed when 5% and 10% H_2S were added to natural gas and also when 5% and 10% natural gas were added to H_2S. The addition of 5% H_2S to natural gas resulted in higher pressure than pure H_2S and pure natural gas. The 5% mixture gave much faster combustion than pure natural gas under fuel rich conditions.

INTRODUCTION

Hydrogen sulfide (H_2S) may be present in various concentrations in crude oil, natural gas, and biogas; an understanding of its effects is necessary since hydrogen sulfide is a toxic, flammable, and corrosive substance. The industrial process of sulfur removal will produce a lot of sulfuric biproducts. These biproducts could be a potential hazard to factory and workers. The mixture of natural gas and hydrogen sulfide has been a safety issue in development of new oil fields recently.

Jianwen et al. [1] described three major releases of hydrogen sulfide and natural gas that caused severe accidents. To reliably calculate the hazardous consequences of a hydrogen sulphide release, knowledge of its properties is critical. Earlier work investigated detonations in hydrogen sulfide, and its laminar properties have also been studied. However, experimental data from H_2S explosions are limited. This work focuses on the turbulent combustion of hydrogen sulfide and summarizes a series of experimental investigations of explosions with H_2S mixtures. These mixtures are composed of pure H_2S, artificial natural gas (NG) (10% propane and 90% methane), and NG mixed with H_2S. All tests are mixed with air and are conducted at 1 atm initial pressure and ambient temperature. A square pipe with repeated obstacles is used to generate turbulence and increase the flame speed in the study. The experimental results provide a reference data set for verification of CFD codes and also enable a direct comparison with natural gas for the maximum pressure. As more unconventional oil sources are developed, there will be an increasing need to accurately model the combustion of natural gas and hydrogen sulfide mixtures for risk assessment.

GAS EXPLOSIONS IN HYDROGEN SULFIDE

Glassman and Yetter [2] provide a general discussion on sulfur combustion which describes the inhibition of oxidation of hydrogen by H_2S. The stoichiometric combustion of H_2S in oxygen can be written as the overall reaction

$$2H_2S + 3O_2 \longrightarrow 2SO_2 + 2H_2O$$

(1).

In a stoichiometric and rich mixture some of the SO_2 products may also react with H_2S to form solid S by the Claus reaction [3]

$$2H_2S + SO_2 \longrightarrow 3S + 2H_2O$$

(2).

Alzueta et al. [4] showed that SO_2 could either promote or inhibit the burning of CO depending on the amount of SO_2 and the stoichiometry. Selim et al. [3] investigated premixed methane-air with added H_2S, and they showed that combustion begins with the thermal and chemical decomposition of H_2S. SO_2 was also found to enhance the dimerization of CH_3 radicals to form longer hydrocarbons. A chemical reaction mechanism of sulfur and hydrocarbons has been proposed by Wendt et al. [5] and Frenklach et al. [6].

Chamberlin and Clarke [7] were early investigators of the laminar flame speed of hydrogen sulfide. Their setup was typical of the period and consisted of a tube that was 1 m long and 2.5 cm in internal diameter. The tube had a burner tip. The maximum flame speed was observed at 10% $(\phi = 0.8)$ and had a value of 0.5 m/s. Also a relatively wide flammable region in H_2S-air mixtures was observed. Kurz [8] used a Bunsen burner method to investigate the effect of a hydrogen sulfide additive on the flame speed of propane, and he also included the flame speed measurements for pure H_2S-air. The flame speed decreased as H_2S was added to the propane, up to the maximum investigated concentration of 6%. However, pure H_2S resulted in a higher flame speed than the mix. A Bunsen flame was also used by Gibbs and Calcote [9] to investigate the effect of the molecular structure on the burning velocity for different equivalence ratios. These three experimental studies of H_2S flame speeds are summarized in Figure 2. As seen, there are relatively large discrepancies between the results, and it is also worth noting that none of the results consider the flame stretch effects. This work does not involve any determination of the laminar flame properties but states that the current knowledge of hydrogen sulfide flames is inconsistent. As such it does not provide a good basis for evaluation of potential hazards as compared to other gases.

There is need for further experimental investigations into the laminar burning velocities and chemical kinetics for pure H_2S gas and H_2S mixed with hydrocarbons. These studies could provide more consistent information regarding the laminar flame properties of the fuel and chemical induction delay times. Such data would be valuable as input to modelling tools and validation of chemical reaction mechanisms. Until new knowledge has been found, one must use the methods available but beware of its limitations.

Cantera software was used to calculate the constant volume combustion pressure and the constant pressure expansion ratio by the reaction mechanism of Wendt et al. [5]. These results are given in Figure 1 and are calculated for stoichiometric fuels, with the H_2S content in NG ranging from 0 (pure natural gas) to pure H_2S, using increasing additions of H_2S. It is shown that the equilibrium pressure and expansion ratio are inversely proportional to the hydrogen sulfide content in the fuel. The calculations suggest that there should be lower flame speed and pressure build-up in propagating hydrogen sulphide deflagration than natural gas mixtures.

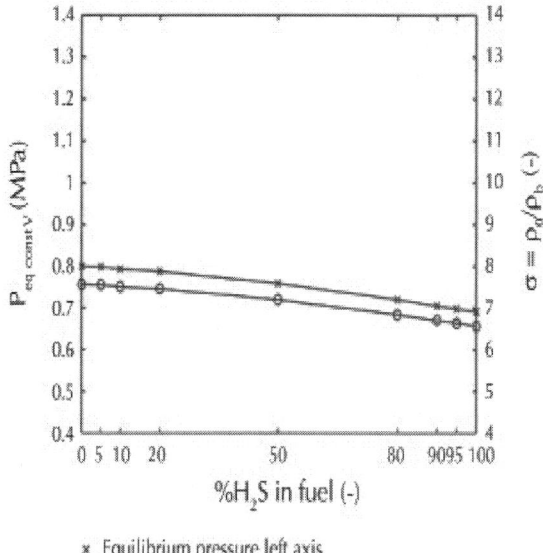

Figure 1: Cantera calculation. Constant volume combustion equilibrium pressure for stoichiometric fuel ranging from pure NG (left) to pure H₂S (right).

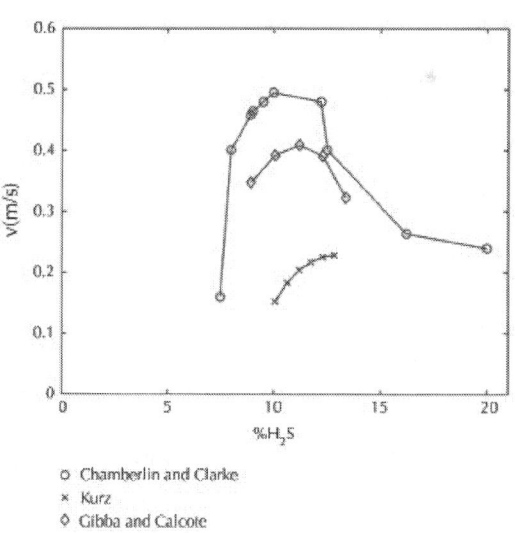

Figure 2: Flame speed of H₂S-air mixtures at different concentrations.

Bozek and Rowe [10] compared fuel properties from the International Electrotechnical Commision (IEC) and the National Fire Protection Association (NFPA). Both datasets show that the flammability region of hydrogen sulfide is wider than that of methane and pentane. Pahl and Holtappels [11] at the BAM Federal Institute for Materials Research and Testing investigated the explosion limits of H_2S and air in mixtures with N_2 and CO_2. They found the upper and lower explosion limits to be 49.8% and 3.9%, respectively. When CO_2 or N_2 was added to the mixture, the measured explosion limits were higher than those found in an earlier work by Coward and Jones [12].

The minimum experimental safe gap MESG for hydrogen sulfide is lower than that for methane and pentane, which indicates the reactivity of the fuel. NFPA 68 Guide for Venting of Deflagrations (2002) provides data for the deflagration index and shows that it is higher for methane than for H_2S.

Moen and coworkers [13–17] investigated flame acceleration and detonations in H_2S mixtures. The detonation cell size of hydrogen sulfide detonations was 100 mm, while those of methane and propane were 280 mm and 69 mm, respectively. This indicates that H_2S mixtures detonate easier than methane. The deflagration to detonation transition (DDT) of H_2S mixtures has not been widely investigated. Moen et al. [16] investigated the flame acceleration of H_2S-air mixtures in a 1.8 m by 1.8 m cross-section and 15.5 m long square pipe, with obstacles made of steel pipes with diameters 500 mm and 220 mm. They compared the results to those using acetylene-air mixtures. For the hydrogen sulfide experiments they recorded overpressures of only 20 to 50 mbar and flame speeds from 36 to 81 m/s. In a comparison to acetylene they suggested that the H_2S-air mixtures could detonate if the scale was large enough, the ignition was strong enough, or sufficient confinement was present.

Shepherd et al. [17] and Vervisch et al. [18] studied the activation energy of hydrogen sulfide and compared it to other fuels. The resultant value was 109.67 kJ/mol in the Shepherd study and 92 kJ/mol in the Vervisch study. Turns [19] gave 125 kJ/mol activation energy for propane and 125 kJ/mol or 202 kJ/mol for methane.

EXPERIMENTAL SETUP

The experimental setup used in this work was made from a stainless steel square pipe with inner dimensions of 84 mm. Four parts were bolted together and sealed to make an airtight compartment. Figure 4 shows a schematic of the four parts with their dimensions, obstacle spacing, and pressure transducer positions. Figure 3 shows a picture and Figure 5 shows a sketch of the assembled setup. The experimental setup was chosen to facilitate strong flame acceleration in the beginning and enough spacing in section 2 to possibly get local volume explosions or DDT. The experimental setup was tested also for propane, methane.

Figure 3: Picture of the experimental setup.

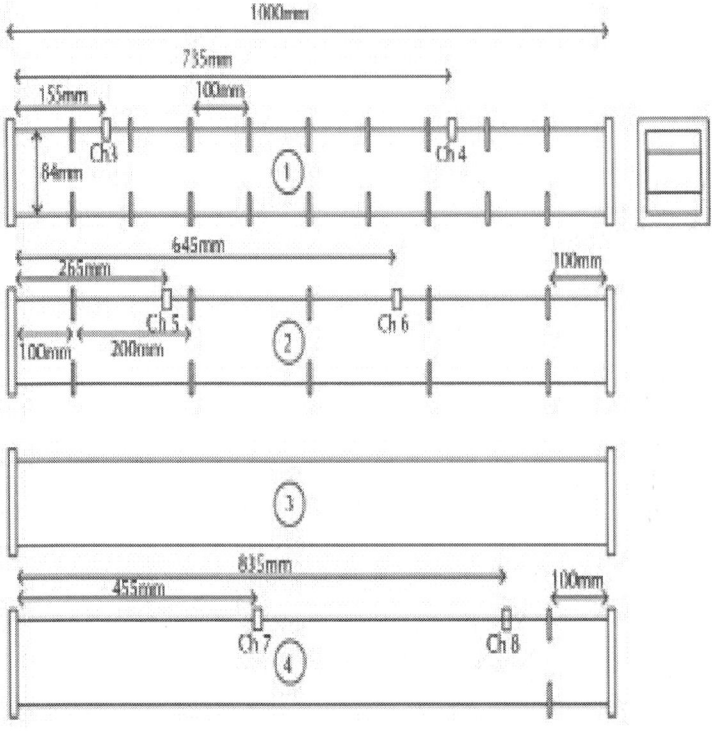

Figure 4: Schematic drawing of the experimental setup. Ch# = pressure recording channel number.

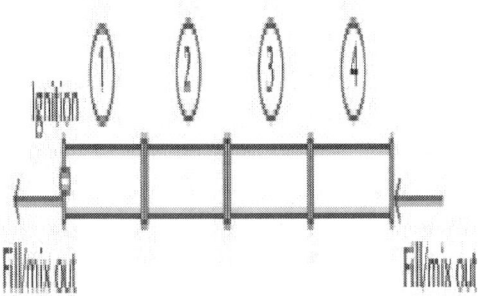

Figure 5: The experimental setup consisting of four stainless steel tube sections.

The pressures were recorded with two Kistler 7001 (Ch 1 and Ch 2) and four Kistler 603b (Ch 3 to Ch 6) piezoelectric transducers (Figure 2) and an oscilloscope recording at 1 MHz. The ignition system was a center-mounted 10 kV spark at the end flange of section 1. At 10 cm from the end of section 4, one obstacle was installed not only to add strength but also to reflect any shock waves and achieve DDT (if possible) at the end obstacle. DDT located at the end flange is undesirable since it would cause strain on the bolts and filling system.

The fuel-air mixture was made by evacuating air from the square pipe and filling it with fuel. All tests were done with 1 atm initial pressure and ambient temperature. A circulation pump was used to circulate and mix the gas through the system. The setup was placed with the obstacles in vertical alignment. This prevented the fuel from being "trapped" in the pockets between the obstacles at the top and bottom of the pipe. The pump and piping was isolated from the setup before ignition.

Special consideration was made regarding the toxicity of hydrogen sulfide and the sulphur dioxide combustion product. A coal filter with special coated coal was installed at the purge of the square pipe to remove sulfuric components from the gas. No H_2S was measured at the outlet of the ventilation system.

This work was part of a larger study to compare H_2S and natural gas mixtures to other more determined fuels. The fuels were acetylene, hydrogen, propane, methane, synthetic natural gas, and H_2S. All fuels were mixed with air. Four different combustion regimes were observed in the study.

To illustrate these explosion regimes, the pressure records are plotted in a diagram showing time along the y-axis and pressure plus the positions of the pressure transducers along the -axis. This type of diagram gives a good display of the trajectory of the pressure waves, shock waves, and detonation waves in a gas explosion. Figure 6 shows these four different explosion regimes in these types of diagrams:

- slow flame propagation and no shock waves formed in front of the flame, which is well known as a slow flame regime;
- fast flame propagation (regime) and shock wave formed but no strong local explosion due to reflection of the shock at the end of the pipe;

- fast flame propagation and shock wave with local explosion and transition to detonation due to reflection of the shock wave at the end of the pipe;

- fast flame propagation and transition to detonation in obstructed area or close to the exit of the obstructed part of the pipe.

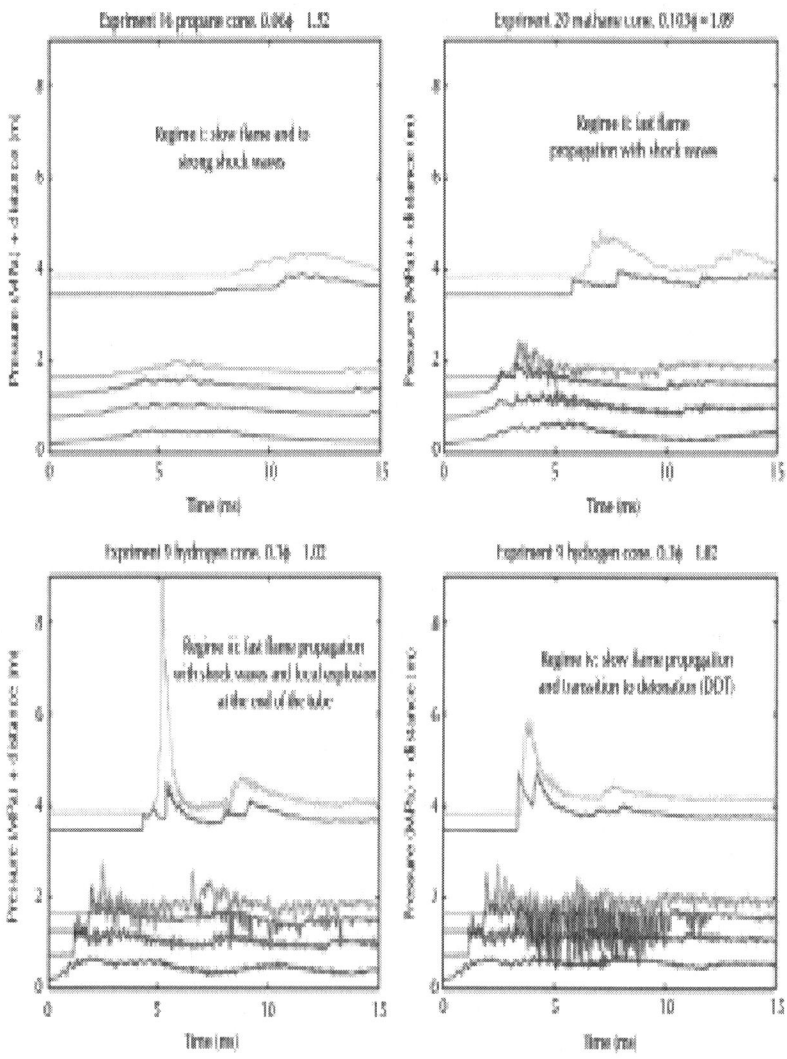

Figure 6: The four different explosion regimes.

Only slow and fast flames were observed in the experiments reported in this paper, but the other regimes are given to provide a qualitative justification of the assumed flame propagation.

Since there is no visual recording of the flame fronts, it is only assumed that the deflagration was similar to other reported works in a very similar setup. Details of this can be found in Lee [20]. The flame fronts become stretched and unstable as they propagate through the obstacles, and the flow through the obstacle openings can enhance the mixing at the flame front. Shock reflections at the solid obstacles are also well known to cause local explosions or DDT in sensitive gas mixtures.

RESULTS

The fuel mixtures used in this work were pure hydrogen sulfide and fuel mixtures with artificial natural gas (premade 10% propane and 90% methane). The experiments with pure natural gas (NG) and pure H_2S in air are presented first to provide a basis for comparison. Next, results from pure H_2S are presented, and last the mixtures of H_2S and NG in air are presented.

The experimental matrix in Table 1 shows the gases, concentrations, and equivalence ratios.

Table 1: Experimental matrix

Test #	Gas 1	Vol. %	Gas 2	Vol. %	ϕ
23	NG	8.30			0.99
24	NG	6.20			0.72
25	NG	9.20			1.11
26	NG	10.40			1.27
27	H2S	10.00			0.79
28	H2S	12.40			1.01
29	H2S	9.00			0.71
30	H2S	15.10			1.27
49	H2S	25.00			2.38

31	H2S	0.43	NG	8.08	1.00
32	H2S	0.32	NG	5.99	0.72
33	H2S	0.53	NG	9.98	1.26
34	H2S	0.86	NG	7.74	0.99
35	H2S	1.80	NG	7.20	1.01
36	H2S	5.00	NG	5.00	1.00
37	H2S	8.96	NG	2.24	1.00
39	H2S	10.53	NG	1.17	1.00
40	H2S	11.40	NG	0.60	1.00

Natural Gas

As reference experiments, tests were conducted using artificial natural gas. The concentrations were 6.2%, 8.3%, 9.2%, and 10.4% corresponding to equivalence ratios $(\phi = 0.72)$ of, 0.99, 1.11, and 1.27.

Pressure records from the stoichiometric experiment are given in Figure 7. The pressure curves are offset along the vertical axis, an amount equal to the distance of the transducer from the ignition end. After ignition the flame first propagated through the obstructed part of the pipe. This caused the flame to increase in surface area, and the flow ahead of the flame became turbulent. The turbulent flow caused the flame to accelerate and increase its reaction rate. This is seen in the pressure plots as the rate of pressure gradient increases. At early times a slow pressure increase was observed on channels 1 and 2, with a faster pressure rise seen on channels 3 and 4. In the smooth section a propagating shock wave was recorded on channel 5 at 5.5 ms. this was generated as the flame accelerated and the displaced flow ahead was fast enough. The shock wave was recorded on channel 6 at 6.2 ms and a reflection at the end obstacle was recorded at 6.5 ms. the reflected shock wave was also recorded on channel 5 at 7.5 ms as it propagated backwards toward the ignition end. Further details on flame acceleration in obstructed pipes can be found in Ciccarelli and Dorofeev [21].

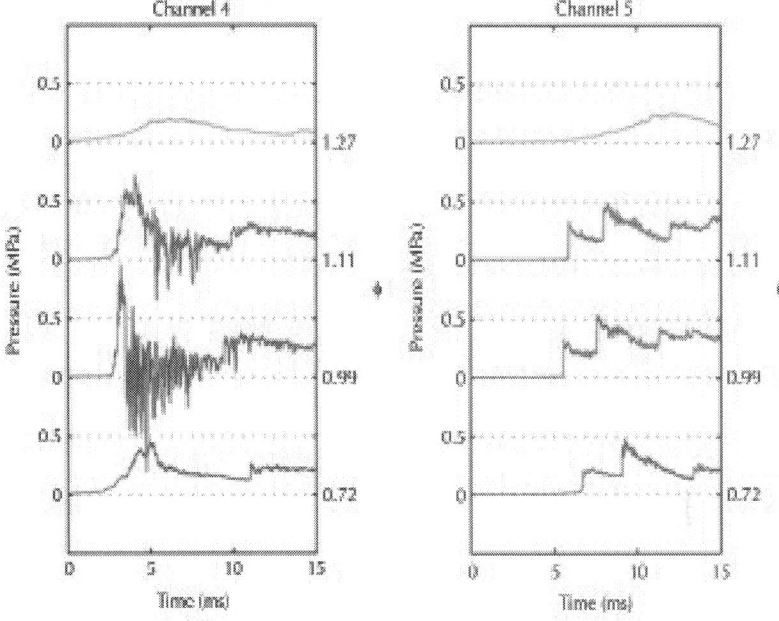

Figure 7: Pressure records from the stoichiometric NG-air mixture (test 23). Channels 1–6 are shown from bottom to top. The pressure levels are offset, an amount equal to the distance (m) from the ignition end.

A comparison plot from the natural gas experiments with different fuel concentrations is given in Figure 8. The pressure is read on the left vertical axis and the equivalence ratio is shown on the right vertical axis. The horizontal axis shows the time. The leanest experiment ($\phi = 0.72$), with 6.2% fuel in air, showed a pressure rise of almost 0.5 MPa in the obstructed part of the experimental setup (channel 4) and a primary pressure wave of about 0.25 MPa in the smooth section. The stoichiometric experiment with 8.3% fuel in air showed the fastest pressure rise and the highest pressure (1 MPa). A 0.3 MPa shock wave was recorded in the smooth section. For 9.2% fuel in air ($\phi = 1.11$), the pressure rise in the obstructed section was lower than that in the stoichiometric experiment, while the shock wave in the smooth section was almost equal. The richest experiment (10.4% fuel in air, $\phi = 1.27$) resulted in a slow flame and a very slow pressure rise recorded on all pressure transducers.

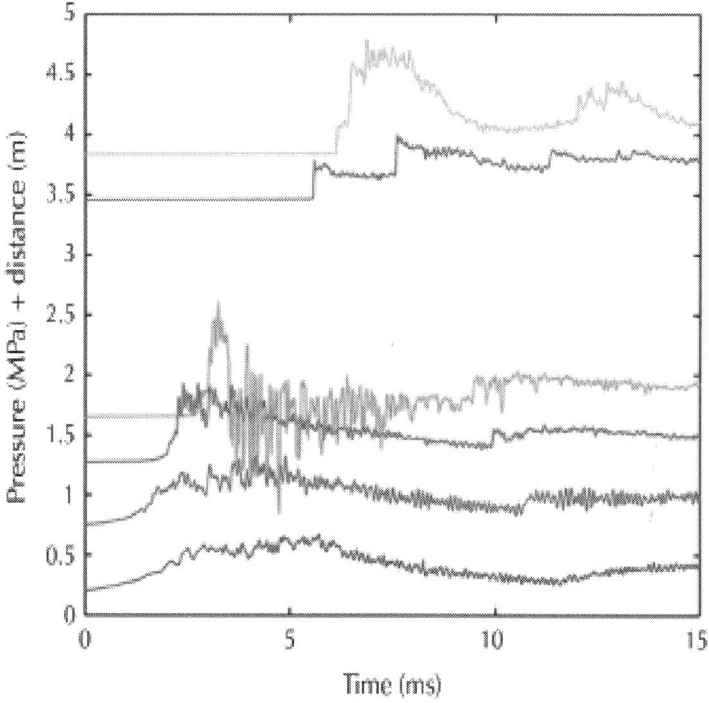

Figure 8: From bottom: tests 24, 23, 25, and 26. Comparison of pressure records from channels 4 and 5 for lean, stoichiometric, and rich NG-air mixtures. Pressure is shown on left vertical axis, while the equivalence ratio is given on the right vertical axis.

Hydrogen Sulfide and Air Mixtures

Results from five tests with the pure H_2S-air mixture are presented. The H_2S concentration ranged from 9% to 25% (see Table 1), where 12.4% is the stoichiometric concentration. The pressure records from the stoichiometric experiment are shown in Figure 9. The overall phenomenon is similar to the stoichiometric natural gas experiment. The initial slow burning and the subsequent development to a faster turbulent flame are seen in the pressure plot. The pressure levels on channels 1 to 4 are lower than in the NG experiment, indicating that this experiment burned slower. The shock wave in the smooth section was roughly the same as in the NG experiment.

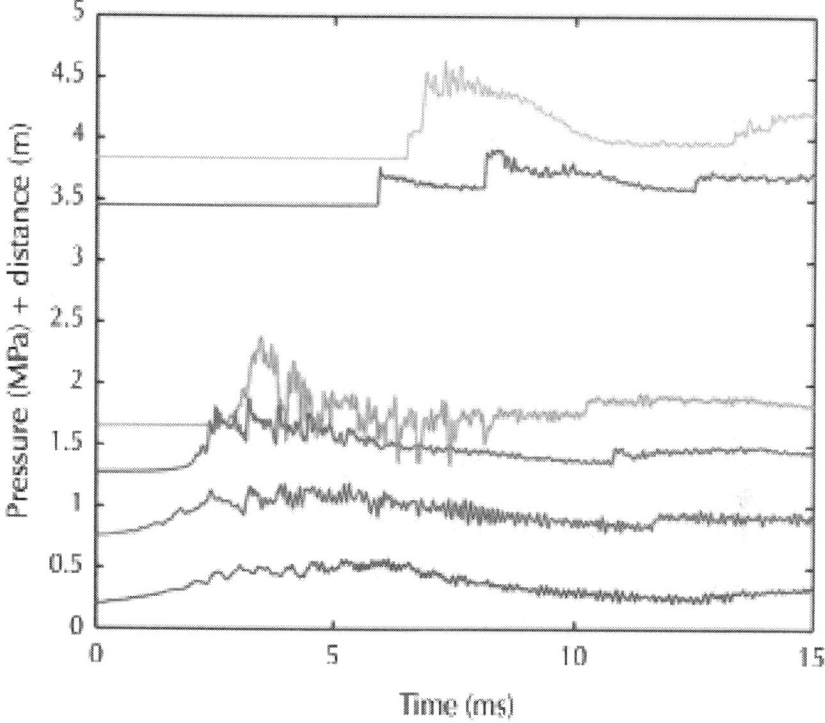

Figure 9: Pressure records from the stoichiometric H_2S-air mixture (test 28). Channels 1–6 are shown from bottom to top. The pressure levels are offset, an amount equal to the distance (m) from the ignition end.

Figure 10 shows a comparison plot of the hydrogen sulfide experiments, with the pressure shown on the left vertical axis and the equivalence ratio shown on the right vertical axis. The horizontal axis shows the time. The leanest mixture was 9% H_2S in air ($\phi = 0.71$) and showed a pressure rise of about 0.3 MPa. It did not result in a shock in the smooth section of the setup. The recorded pressure wave was about 0.2 MPa, and it reflected at the end wall and obstacle. The slightly richer mixture of 10% ($\phi = 0.79$) showed a 0.3 MPa shock wave propagating in the smooth section of the experimental setup. In the obstructed part, 0.5 MPa was recorded at channel 4.

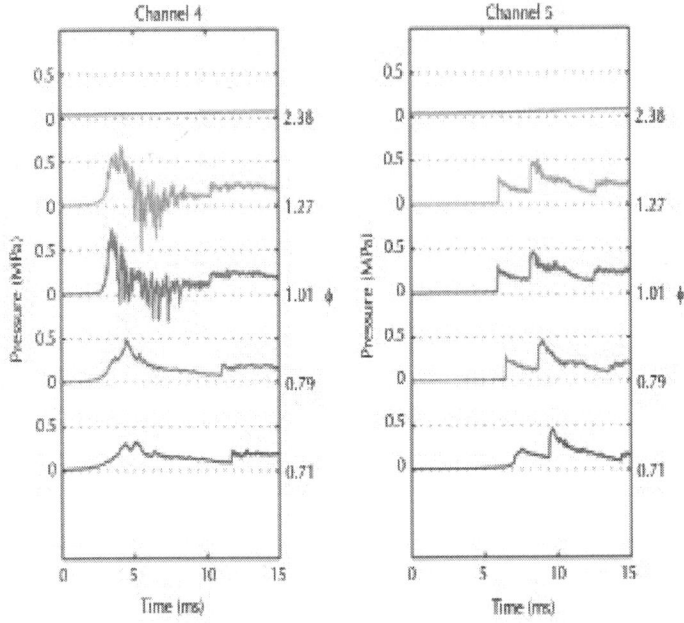

Figure 10: From bottom: tests 29, 27, 28, 30, and 49. Comparison of pressure records from channels 4 and 5 for lean, stoichiometric, and rich H_2S-air mixtures. Pressure is shown on left vertical axis, while the equivalence ratio is given on the right vertical axis.

The stoichiometric mixture resulted in a 0.35 MPa shock in the smooth section, while 0.75 MPa was recorded in the obstructed section.

The experiment with $(\phi = 1.27)$ corresponding to 15.1% H_2S in air was very similar to the stoichiometric case, with only 0.05 MPa lower pressure in the smooth section and the obstructed section. Due to the wide flammability region of H_2S, $(\phi = 2.38)$ was also investigated; it resulted in a very slow flame and a low pressure increase of about 0.1 MPa.

H_2S-Natural Gas-Air Experiments, Results, and Discussion

Experiments were performed on a set of nine tests, with the first three containing 5% H_2S and 95% natural gas. The equivalence ratios were

$\phi = 0.72$, $\phi = 1.00$, and $\phi = 1.26$ The following experiments were all conducted with but with increasing hydrogen sulfide content. The H$_2$S fractions in natural gas were 5, 10, 20, 50, 80, 90, and 95%.

Figure 11 shows that, by keeping the H$_2$S-to-NG ratio constant at 5:95 and varying the equivalence ratio, $\phi = 0.72$ and $\phi = 1.26$ give quite similar pressure levels: 0.5 MPa in the obstructed part and 0.3 MPa in the smooth section. The stoichiometric experiment resulted in the fastest pressure rise and a peak pressure of more than 1.3 MPa. A shock wave of 0.4 MPa was recorded in the smooth section. The rich mixture ($\phi = 1.26$) resulted in strong flame acceleration, 0.5 MPa recorded on channel 4, and a pressure wave in the smooth section.

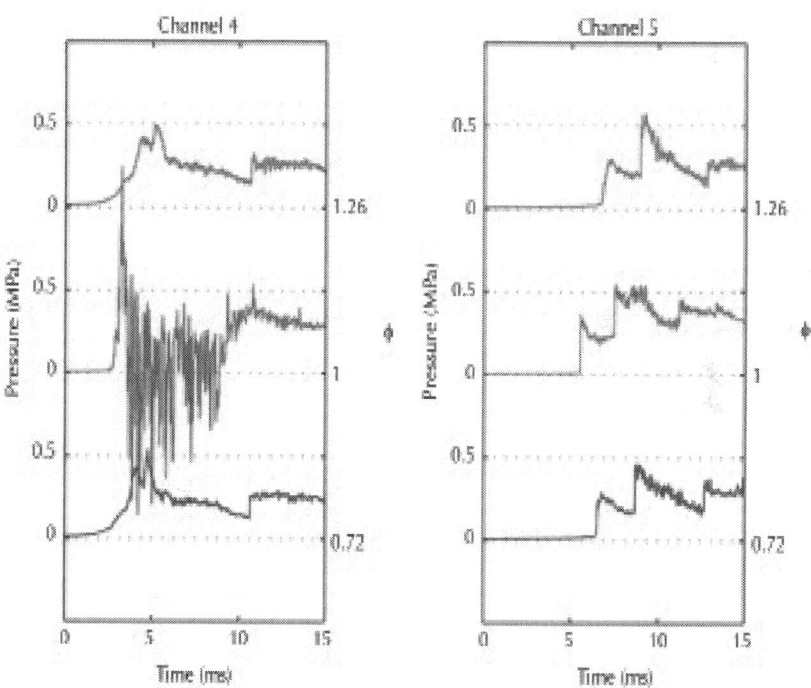

Figure 11: From bottom: tests 32, 31, and 33. Comparison of pressure records from channels 4 and 5. Lean, stoichiometric, and rich 5% H$_2$S/95% NG-air mixtures. Pressure is shown on the left vertical axis, while the equivalence ratio is given on the right vertical axis.

With the equivalence ratio kept constant at 1 and the H_2S content in the fuel varied from 0% to 100%, the pressure did not change much except for some spikes, as seen in Figure 12. The pressure is shown on the left vertical axis, and the H_2S content in the fuel is shown on the right vertical axis. Time is shown on the horizontal axis. The pressure in the obstructed part was recorded between 0.8 and 1 MPa, and the shock propagating in the smooth section was about 0.3 to 0.35 MPa and reflected at 0.5 MPa.

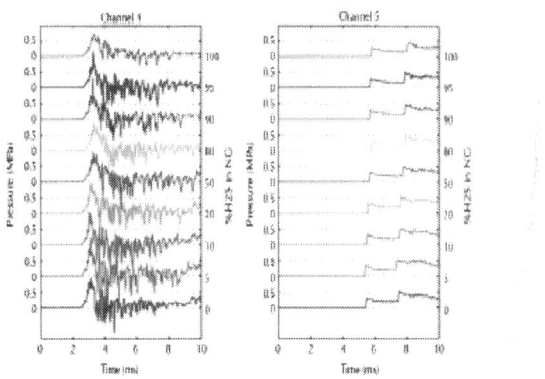

Figure 12: From bottom: tests 23, 31, 34, 35, 36, 37, 39, 40, and 28. Comparison of pressure records from channels 4 and 5. The mixture varies from pure natural gas in the fuel (bottom) to pure H_2S in the fuel (top). All experiments are stoichiometric mixtures.

DISCUSSION

The experimental study for pure natural gas and air showed that the flame propagated fast when the equivalence ratio was lower than 1.27, producing strong deflagrations in the experimental setup. The pressure results showed that the rate of energy release increased as the flame propagated through the square pipe. The richest natural gas mixture investigated was $(\phi = 1.27)$, and that mixture resulted in a slow pressure rise believed to be due to a slow burning velocity of the flame.

The explosion pressures for lean H_2S-air were slightly lower than the pressures for lean NG-air. The lower explosion pressures were to some extent a result of the lower expansion ratio of the H_2S-air flame compared with the other fuels. The expansion ratio $(\sigma = \rho_u / \rho_b)$ of H_2S is about 6.6 while it is 7.6 for NG. This results in a lower flame speed, less turbulence, and, therefore, a lower pressure rise.

By comparing the H_2S-air mixtures with mixtures of natural gas and air, as shown in Figure 10 and Figure 8, it was observed experimentally that natural gas and H_2S result in a fast flame for $(\phi = 0.72)$. On the rich side $(\phi = 1.27)$, the hydrogen sulfide accelerated as a fast flame while the natural gas was slow. This was expected due to the wider flammability region of H_2S [10] compared with NG.

The experiments with stoichiometric H_2S-NG-air showed that the flame in the experimental setup produced strong deflagrations with high pressures in the obstructed part of the experimental setup. The pressures seen with channel 4 in tests with 90% and 95% H_2S in the fuel (1.35 and 1.15 MPa) indicate that the compression heating of the reactants caused local ignition in a hot spot.

Comparing the maximum pressure from channels 2, 4, and 5, a trend is observed in Figure 13 in which the maximum pressure decreases as the H_2S content in the fuel increases; however, the spikes are also observed when plotting the maximum pressure for three channels when the hydrogen sulfide content was varied. These spikes correspond to 90% and 95% H_2S in the fuel as well as 5% and 10% H_2S in the fuel.

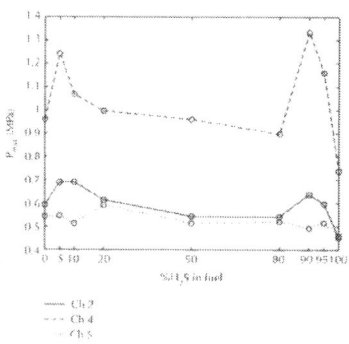

Figure 13: Maximum pressure from experiments. The pressure from channels 2, 4, and 5 for various H_2S contents in the fuel.

Compared to the constant volume and constant pressure calculations in Figure 1 it is clear that the pressure spikes originate from different phenomena. One possible explanation could be a more sensitive mixture when small amounts of H_2S are added to natural gas or the opposite. A reduction in chemical induction delay time could lead to local explosions in heated volumes of reactants. These local explosions are very hard to determine even with full view of the channel, but other studies have shown that they are more likely to occur in the obstructed part rather than in the unobstructed parts (Lee [20]).

By comparing Figures 13 and 1, it can be seen that the pressure on channel 4 (section with obstacles) exceeds the constant volume pressure. The equilibrium pressure and the expansion ratio do not explain the spikes seen in Figure 13.

Hot spots and local ignition are closely related to deflagration to detonation transition (DDT), which results in high pressure. No DDT was recorded in these experiments, but the pressure spikes suggest that local explosions could have occurred.

There are always uncertainties when reporting the maximum pressure, since it is measured at one position. Other spikes that may occur in other sections of the experimental setup may be missed by the transducer recording.

By keeping the H_2S content in the fuel constant and changing the equivalence ratio, differences are observed in the combustion. Figure 14 shows the maximum pressure results from the tests with 100% NG, 100% H_2S, and 5% H_2S in NG (mix) for different equivalence ratios.

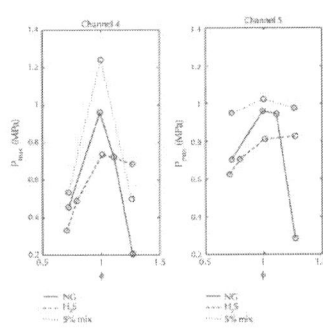

Figure 14: Maximum pressure for different equivalence ratios, for pure NG, pure H_2S, and 5% H_2S in NG.

The addition of 5% H₂S to the natural gas makes the mixture more reactive and, therefore, results in a higher pressure than that with pure NG and pure H₂S. Another notable effect is that the mixture becomes much more insensitive to changes in the equivalence ratio when comparing the maximum pressure from channel 5; that is, it produces higher pressure on both lean and rich sides compared with pure fuels.

A comparison of the pressure in the obstructed section and the smooth section with and without 5% hydrogen sulfide is shown in Figures 15 and 16. Figure 15 shows the stoichiometric case, and the two pressure records from channel 4 and the two pressure records from channel 5 have the same shape and order. This indicates a similar combustion process.

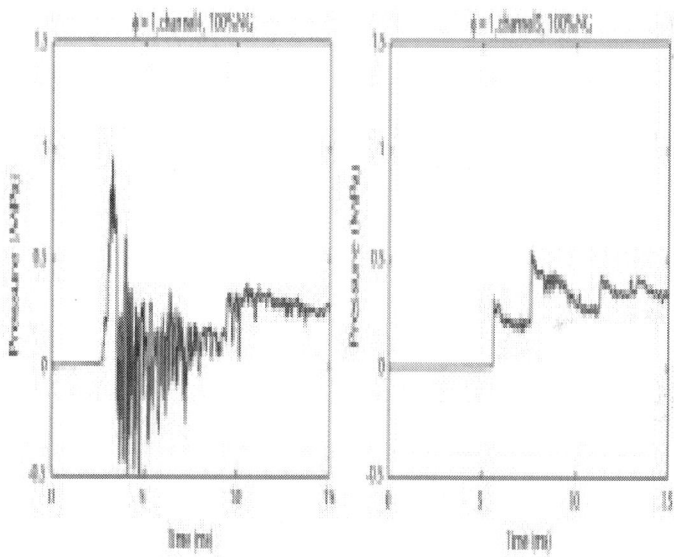

(a)

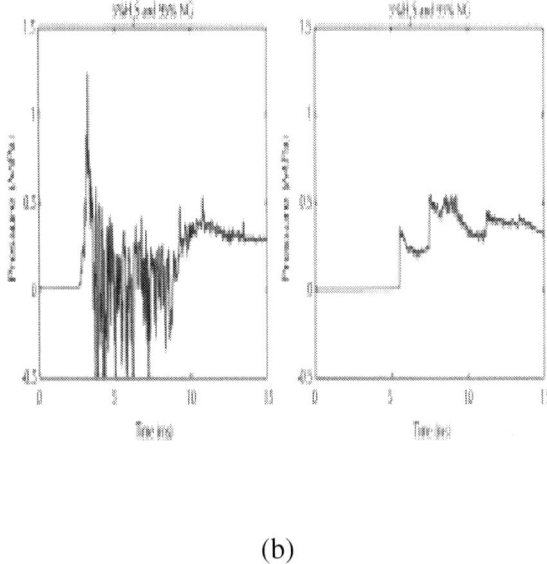

(b)

Figure 15: Comparison of explosion pressures for $(\phi = 1)$ in the obstructed section (channel 4) and the smooth section (channel 5). (a) 100% NG and (b) NG with 5% H_2S.

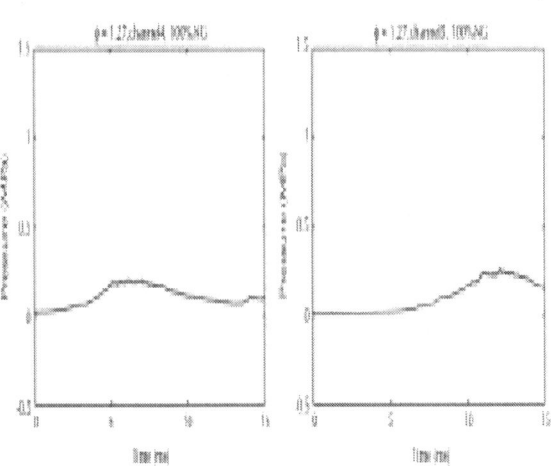

(a)

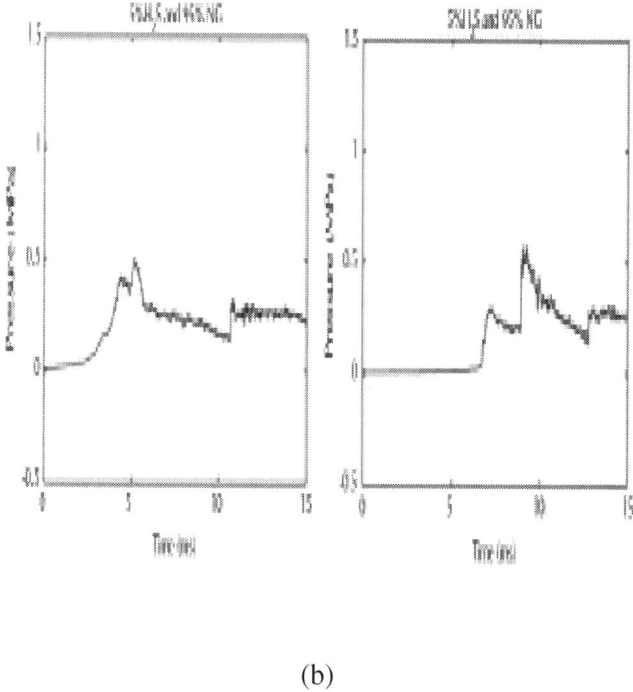

(b)

Figure 16: Comparison of explosion pressures for ($\phi = 1.27$) in the obstructed section (channel 4) and the smooth section (channel 5). (a) 100% NG and (b) NG with 5% H$_2$S.

When comparing the explosion pressures with the rich cases (Figure 16), it is seen that there is a major change in the pressure recordings when comparing the same channel. The pure NG burns slowly (a), while the mixed fuel (b) burns much faster and results in a strong pressure wave in the smooth section. This is a significant change caused by the addition of relatively small amounts of hydrogen sulfide to the fuel. There is still more to investigate regarding the combustion of hydrocarbons and sulfur compounds.

These experiments are small/medium scale, and the scale effects of hydrogen sulfide and natural gas explosions are still unknown; however, the presence of hot spots and pressure spikes suggests that DDT might occur if the scale was larger. It was suggested by Moen [15] that the use of a denser obstacle field in experiments would increase the turbulence and flame speed.

CONCLUSIONS

Only limited data are available in the open literature on H_2S-air deflagrations and especially H_2S and natural gas mixtures. Data for explosions at conditions supporting strong flame acceleration are lacking. In the present work we have successfully performed such experiments and obtained new and unique experimental data for explosions with hydrogen sulfide and natural gas mixtures. A comparison to pure natural gas is also included.

- Pure fuels: hydrogen sulfide has a wide flammability region compared with methane and propane, as shown in the literature. In this study, H_2S-air mixtures produced lower explosion pressures at lean-to-stoichiometric compositions relative to natural gas. On the rich side, the H_2S-air mixtures produced higher explosion pressures.

- Fuel mixtures at $\phi = 1$ a decrease in the maximum pressure was observed when increasing amounts of hydrogen sulfide were added to the natural gas. There were, however, some maximum pressure spikes observed for 90% and 95% H_2S in NG, as well as for 5% and 10% H_2S in NG. These spikes could be a result of a local explosion of compressed reactants, but they did not develop into detonations.

- Rich fuel mixtures: rich NG with 5% hydrogen sulfide is more reactive than pure rich NG. When 5% H_2S was added to the NG at $\phi = 1$, the result was similar to pure NG but with spikes. When the stoichiometry was changed to $\phi = 1.27$ the result was a fast flame and a strong pressure wave formation in the 5% mixture, while the pure NG had a slow deflagration and a slow and low pressure rise. The 5% mixed fuel also showed decreased sensitivity to changes in the equivalence ratio when the maximum pressures from channel 5 were investigated. These results are important to the process and petroleum industry.

For further work, it is suggested that the experimental results are compared to numerical simulations using commercial and academic software. There is also a need for a thorough study of the laminar properties of H_2S-hydrocarbon-air mixtures. Further experimental investigations should be conducted with higher and lower blockage ratios. Larger scale experiments could reveal the possibility of DDT in H_2S mixtures, and investigations of rich mixtures should be conducted to better understand the effects of added hydrogen sulfide to natural gas.

ACKNOWLEDGMENTS

The authors gratefully acknowledge the financial support by Statoil ASA.

REFERENCES

1.   Z. Jianwen, L. Da, and F. Wenxing, "Analysis of chemical disasters caused by release of hydrogen sulfide-bearing natural gas," Procedia Engineering, vol. 26, pp. 1878–1890, 2011.

2.   I. Glassman and R. Yetter, Combustion, Academic Press, 4th edition, 2008.

3.   H. Selim, A. Al Shoaibi, and A. K. Gupta, "Effect of H_2S in methane/air flames on sulfur chemistry and products speciation," Applied Energy, vol. 88, no. 8, pp. 2593–2600, 2011

4.   M. U. Alzueta, R. Bilbao, and P. Glarborg, "Inhibition and sensitization of fuel oxidation by SO_2,"Combustion and Flame, vol. 127, no. 4, pp. 2234–2251, 2001.

5.   J. O. L. Wendt, E. C. Wootan, and T. L. Corley, "Postflame behavior of nitrogenous species in the presence of fuel sulfur I. Rich, moist, $CO/Ar/O_2$ flames," Combustion and Flame, vol. 49, no. 1–3, pp. 261–274, 1983.

6.   M. Frenklach, J. H. Lee, J. N. White, and W. C. Gardiner Jr., "Oxidation of hydrogen sulfide,"Combustion and Flame, vol. 41, pp. 1–16, 1981.

7.   D. S. Chamberlin and D. R. Clarke, "Flame speed of hydrogen sulfide," Proceedings of the Symposium on Combustion, vol. 1-2, no. C, pp. 33–35, 1948.

8.   P. F. Kurz, "Influence of hydrogen sulfide on flame speed of propane-air mixtures," Industrial & Engineering Chemistry, vol. 45, no. 10, pp. 2361–2366, 1953.

9.   G. J. Gibbs and H. F. Calcote, "Effect of molecular structure on burning velocity," Journal of Chemical and Engineering Data, vol. 4, no. 3, pp. 226–237, 1959.

10. A. P. Bozek and V. Rowe, "Flammable mixture analysis for hazardous area classification," inProceedings of the 55th IEEE Petroleum and Chemical Industry Technical Conference (PCIC '08), pp. 1–10, September 2008.

11. R. Pahl and K. Holtappels, "Explosions limits of H_2S/CO_2/air and H_2S/N_2/air," Chemical Engineering & Technology, vol. 28, no. 7, pp. 746–749, 2005.

12. H. Coward and G. Jones, Limits of Flammability of Gases and Vapors: Bulletin 503, US Bureau of Mines, Juneau, Alaska, USA, 1952.

13. A. Sulmistras, I. O. Moen, and A. J. Saber, "Detonations in hydrogen sulphide-air clouds," Suffield Memorandum 1140, Defence Research Establishment Suffield, Alberta, Canada, 1985.

14. A. J. Saber, A. Sulmistras, I. O. Moen, and P. A. Thibault, "Investigation of the explosion hazard of hydrogen sulphide (Phase I)," Research Report, Defence Research Establishment Suffield, Alberta, Canada, 1985.

15. I. O. Moen, "Investigation of the explosion hazard of hydrogen sulphide (phase II)," Research Report, Defence Research Establishment Suffield, Alberta, Canada, 1986.

16. I. O. Moen, A. Sulmistras, B. H. Hjertager, and J. R. Bakke, "Turbulent flame propagation and transition to detonation in large fuel-air clouds," Symposium (International) on Combustion, vol. 21, no. 1, pp. 1617–1627, 1988.

17. J. E. Shepherd, A. Sulmistras, A. J. Saber, and I. O. Moen, "Chemical kinetics and cellular structure of detonations in hydrogen sulfide and air," in Proceedings of the 10th International Committee on the Dynamics of Explosions and Reactive Systems (ICDERS '85), p. 294, Berkeley, Calif, USA, 1985.

18. L. Vervisch, B. Labegorre, and J. Réveillon, "Hydrogen-sulphur oxy-flame analysis and single-step flame tabulated chemistry," Fuel, vol. 83, no. 4-5, pp. 605–614, 2004.

19. S. R. Turns, An Introduction to Combustion, McGraw-Hill, New York, NY, USA, 2nd edition, 2000.

20. J. H. S. Lee, The Detonation Phenomena, Cambridge University Press, New York, NY, USA, 1st edition, 2008.

21. G. Ciccarelli and S. Dorofeev, "Flame acceleration and transition to detonation in ducts," Progress in Energy and Combustion Science, vol. 34, no. 4, pp. 499–550, 2008.

Chapter 6

Fundamentals of Lubricants and Lubrication

Walter Holweger[1]

[1]Schaeffler Technologies AG & Co.KG, R&D Central Materials, Germany

INTRODUCTION

Literature about lubricants is available in all public domains. Readers should search at those platforms in the case of special interests. Citations given here do not represent the full scale but reflect an overview from a today's perspective. [1-7]

Part of this chapter will be the basic chemical structure of lubricants including some property descriptions. Since literature in tribology is innumerous, the reader should check his special area of interest.

Lubricants play a key role in machinery element safety. Their main tasks are

- to keep moving parts apart from each other,
- to take heat out of the contact by their through pass,
- to keep surfaces clean,
- to transport functional additives toward the surface and
- to transfer power in the application (hydraulic, automatic transmission, breaks). [6, 8]

Functionality of lubricants is defined by their chemical structure and their physical properties. Basics of lubrication are covered by organic chemistry to a major and inorganic chemistry to a minor extent. [2, 3]

Lubricants are regulated internationally and locally, e. g. by ASTM (American Standard of Testing Materials) or DIN (Deutsche Industrienorm). Regulation covers the physical, chemical and toxicological description of lubricants including safety guide lines and others. [2, 3]

SOME BASICS

The spatial structure of carbon chemistry defines all activities of the lubricants derived from them. The spatial structure of organic carbon chemicals is given by the binding state of carbon. [10]

Three main types are discussed. Two are essential for lubrication: single and double bonds.

Single Bonds: Tetrahedral Binding

In the tetrahedral binding state, reflecting the status of single bonds, carbon is placed in the centre of a pyramid with bindings into space from the centre to the corner (Figure 1).

Figure 1: Tetrahedral binding of carbon.

Carbon is placed in the centre of the tetrahedral with four attached valences. Within chemical convention in order to abbreviate the structure denotation the atom symbols are neglected.

Carbon may bind to another one by corner to corner. (Figure 2)

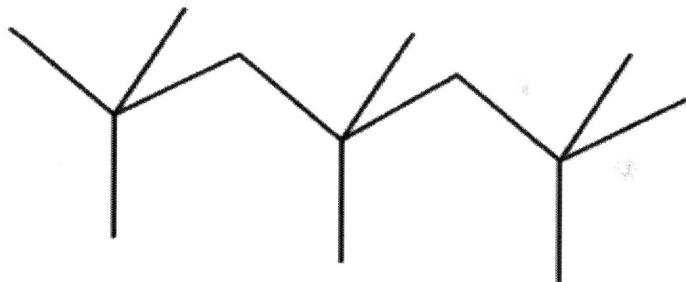

Figure 2: Corner to corner binding state.

Corner to corner binding leads to zigzag chains, where the angle of carbon to the hydrogen atoms is 108°. In general the hydrogen is neglected, leading to a skeleton drawing of the structure.

Beyond the fixed angle of 108° and the zigzag shape of such hydrocarbon structures, a high variety of structures arise due to the fact that those bindings may branch or bind to cyclic structures. (Figure 3and Figure 4)

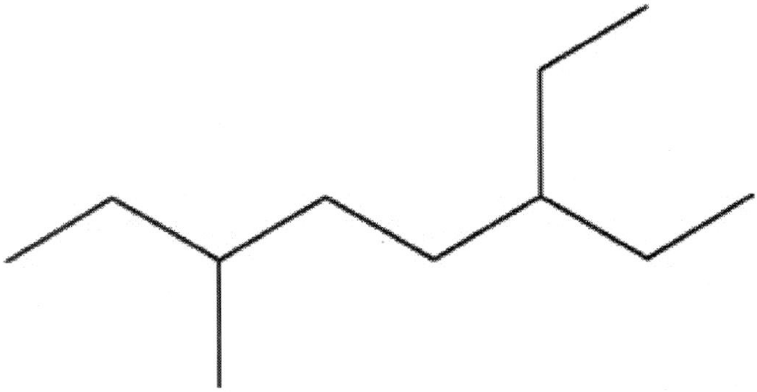

Figure 3: Branched structures by carbon to carbon binding.

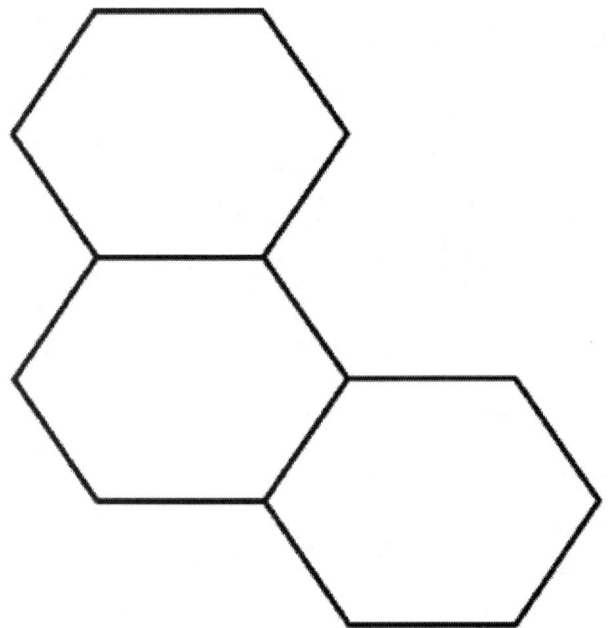

Figure 4: Cyclic Structures by carbon-carbon binding.

Single bonds in hydrocarbons are free to rotate (Figure 5). Rotation leads to the situation that hydrogen atoms within the chain get close to each other. As a consequence the energy of the molecule rises.

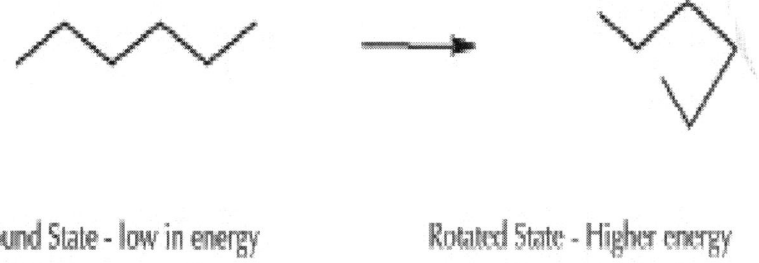

Ground State - low in energy   Rotated State - Higher energy

Figure 5: Energy rise in rotated structures.

Similar to internal rotation, molecular energy rises if molecules get under stress by moving them closely together without giving time to relax. (Figure 6)

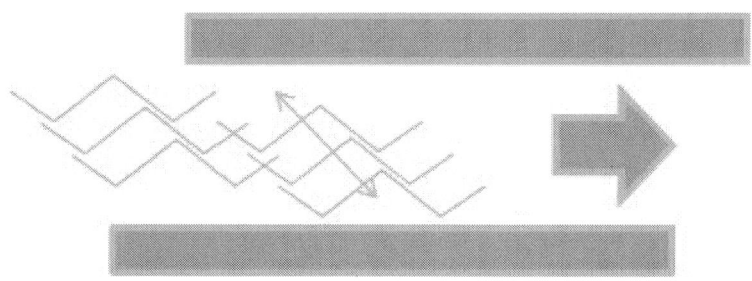

Figure 6: Excitation by pushing molecules to one another by shear stress.

Also the fact of putting or pressing molecules toward a surface may lead to a steep increase in internal molecular energy, sometimes high enough to cut them.

Double Bonds

Carbon may also bind to others by double bonds, such that two of the four bindings attach to the other as double, whereas the remaining bonding stays single. (Figure 7)

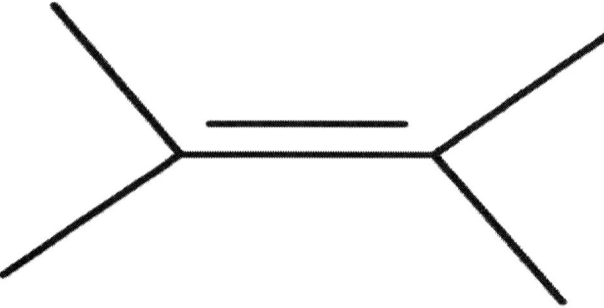

Figure 7: Double bonding.

Double bond shows a 120° neighborhood angle to the carbon. This angle is kept constant and will lead toward different structures in the double bond chemistry. (Figure 8)

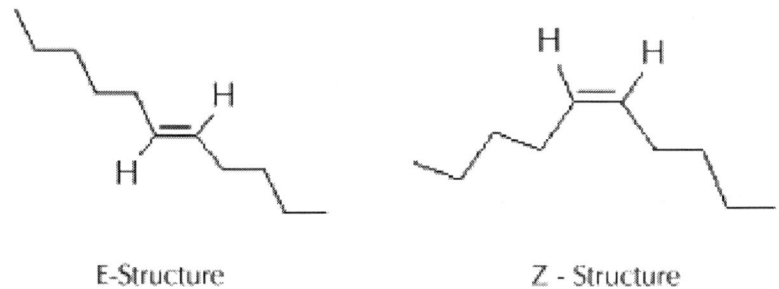

E-Structure   Z - Structure

Figure 8: E and Z structures in double bond.

Both structures differ in their energy. Double bonds are part of biodegradable additives (native oils) but also additives and thickeners in the case of greases. Z-Structures are dominant in native oils.

Triple Bonds

Triple bonds are seldom found in tribology. They represent a high energy state in molecules with very high reactivity. As such, they are part of catalytic degradation processes in lubricants. Within a triple bond carbons attach to each other by a linear principle (Figure 9):

Figure 9: Triple bond present in a hydrocarbon.

BASE OILS IN LUBRICATION: GENERAL COMMENTS ABOUT SPECIE AND GROUPS

Hydrocarbon Base Oils for Lubrication derive from organic chemistry. Different categories are given by their chemical composition and structure. [2]

Hydrocarbons, e.g. Structures that contain solely Hydrogen and Carbon (H, C)

Ester Oils, e.g. Structures containing Hydrogen, Carbon and Oxygen. Some Esters are derived from other precursors, such as phosphoric acid esters.

Polyglycoles: Structure containing Hydrogen, Carbon and Oxygen but being different in binding state compared to Esters.

Within a general scheme, base oils are identified as Groups.

Group I: Those lubricants are built from saturated hydrocarbons, e.g. hydrocarbons without alkenes (hydrocarbons with double bonds)) (> 90%), obtained by solvent extraction processes and catalytic hydrogenation. Sulfur may part in amount of > 0.03%. Viscosity index (VI) is in between 80 and 120.

Group I+: Oils that are in a VI range of 103-108.

Group II: Hydrogenated (saturated) hydrocarbons (> 90%) and sulfur below 0.03% per weight with viscosity index (VI) of 80 till 120.

Group II+: Oils in the VI range of 113-119.

The base oil within this group is manufactured by hydrocracking, solvent extraction or catalytic dewaxing processes. Those oils are pale or water like colored.

Group III: Oils with a saturation > 90%, sulfur < 0.03% and a viscosity index > 120. Those oils are produced by catalytic procedures with a concurrent rearrangement of the carbon backbone during hydrogenation.

Group III+: Oils providing a VI at least of 140.

Group IV: Poly- -Olefins with sulfur content approximately 0, viscosity index 140–170, being produced by catalytic polymerization of low molecular weight end terminated olefins.

Group V: All other oils, e.g. esters, polyglycoles, phosphate esters.

North America states Group III, IV and V as synthesized hydrocarbons (SHC) while in Europe Group IV and V is declared as synthetic oil.

SATURATED NATURAL HYDROCARBONS

Saturated hydrocarbons are those who do not contain double bonds in their structure. They derive from the tetrahedral binding of carbon (bindings that point into corners of tetrahedron). The simplest structure is given by methane, ethane, propane, butane with carbons attached at the corners of the tetrahedral. These representatives are present in the natural gases, while methane is found in enormous quantities as methane-ice cluster. The gases themselves are not in use as lubricants but are components of fuels. (Figure 10)

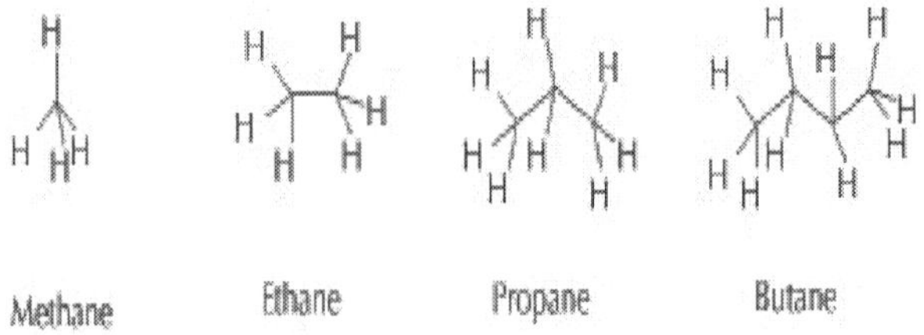

Methane   Ethane   Propane   Butane

Figure 10: Methane, Ethane, Propane, Butane.

Starting from pentane the hydrocarbons get liquid and are the principal components of fuels, solvents, and raw materials for the chemical industry. To facilitate reading and drawing only the carbon backbone is drawn without explicitly showing hydrogen. (Figure 11)

Figure 11: Pentane, Hexane, Heptane.

Binding of carbon to carbon may be realized in chains, but also in branched chains and different cycles (Figure 12).

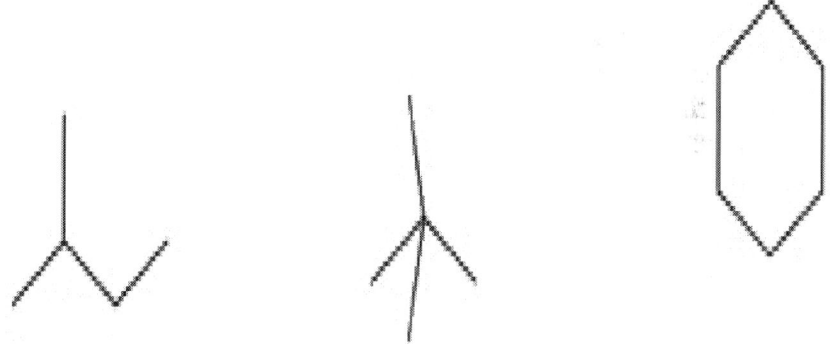

Figure 12: Methylbutane (Isopentane), 2, 2-Dimethylpropane (Neopentane), Cyclohexane

Hydrocarbons from C10 on till C14 are in use as solvents for cleaning (C11-C13 isoparaffines) (Figure 13).

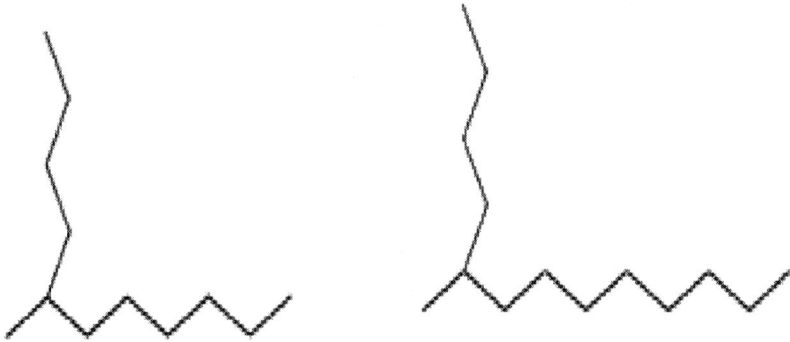

Figure 13: C11-C13 Iso paraffines.

From C16 on, hydrocarbons represent typical structures present in lubricants. As the linear hydrocarbons, beginning at C18 are solids, they are common in waxes and thickeners for liquid hydrocarbons. Due to their high solidification point they are a threat if present in Diesel fuels by blocking filters.

Apart from their function as hydrocarbon waxes they are not suitable as lubricants for machine oil circuits.

Suitable lubricants are derived from C16–C70 hydrocarbons with branched chains. Branching leads to low pour points (the point where the lubricant starts to get solid). Machine oils with low pour point, suitable for low temperature applications are branched in their carbon chain. (Figure 14)

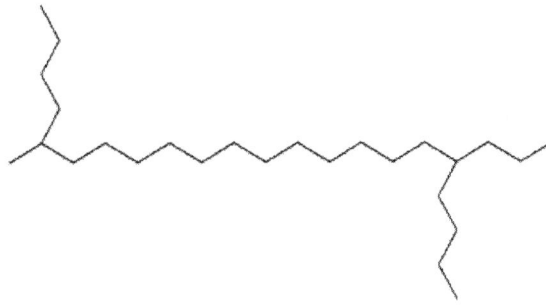

Figure 14: Representatives of saturated hydrocarbons as typical lubricants.

In general, the viscosity of a lubricant - as a measure for the ability to move across - increases with the molecular weight, e.g. the number of carbon atoms attached. Viscosity is measured by different techniques. Basically the lubricant is pushed or moved in between plates or by moving it in the gravity field. International convention states 16 classes of viscosity as an ISO Standard (ISO VG classes): ISO VG 5, 7, 10, 15, 22, 32, 46, 68, 100, 150, 220, 320, 460, 680, 1000 and 1500. Low numbers indicate low viscosity, higher numbers high viscosity. Since viscosity is strictly related to temperature, the ISO VG classification refers to 40°C as a standard temperature. The nature of measuring the viscosity leads to the physical value of an area per time: mm^2/s. Hence, ISO VG 68 for example denotes a viscosity of the lubricant, measured at 40°C within 68 mm^2/s within a range of roughly 10% below and 10% beyond the given 68mm^2/s.

Low molecular weight, branched hydrocarbons are often used in *pneumatic spraying*, due to their viscosity range, starting at 2 (water-like), 5, 10 and 15.

Low viscous hydrocarbons from ISO VG 10, 15, 22, 32, 46, 68 and 100 are in use as *hydraulic oils.*Common hydraulic oil viscosity is around ISO VG 32, 46 and 68.

Hydrocarbons with higher viscosities are part of *machine oils*, carrying out the ordinary lubrication functions. Machine oil viscosities are in the range of ISO VG 68, 100, 150, 220, 320, 460. The number of carbons is in the range of 30–80 in the chain.

Some applications in heavy duty processes demand viscosities even higher in the range of ISO VG 680, 1000 and 1500.

Cyclic Hydrocarbons (Naphtenes)

Naphtenic hydrocarbons are derived from hydrocarbon cycles with more or less long chains attached to the cycle. Due to their high branching they are very common in low temperature applications (below -30°C) for hydraulics; low temperature greases. (Figure 15)

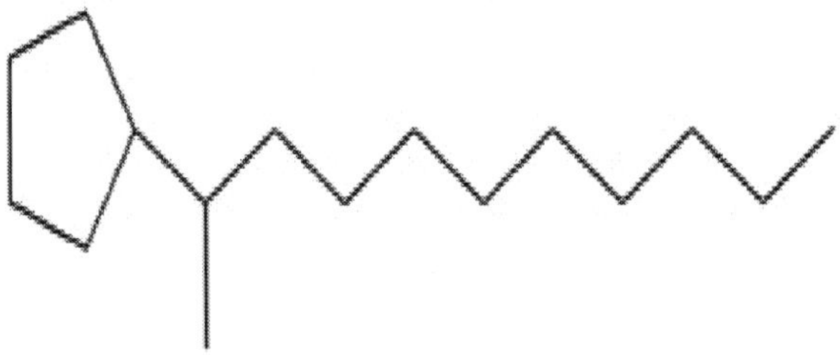

Figure 15: Principal Structure of Napthenic Hydrocarbons.

Aromatic Hydrocarbons (Alkyl Aromats)

Aromatic Hydrocarbons (Alkyl Aromats) derive from the six-membered benzene ring system, attached by hydrocarbon chains. Aromatic hydrocarbons are in use for low temperature applications.

Alkyl Naphthalenes are a modern group of aromatic hydrocarbons. They may act as solvent improvers for synthetic oils, facilitators in generating greases, low temperature applications and much more (Figure 16):

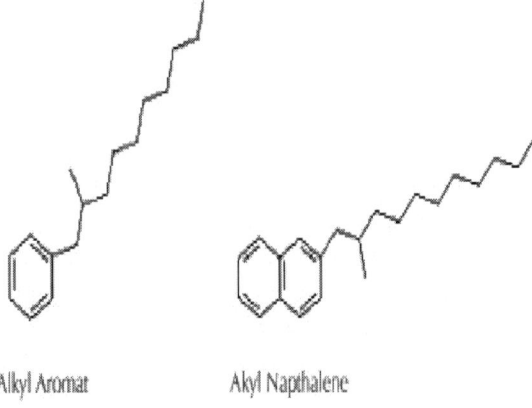

Alkyl Aromat   Alkyl Napthalene

Figure 16: Alkyl Aromats and Naphtalenes.

Aromats and aromat-containing hydrocarbons are very vulnerable toward oxygen. Oxidation of aromats starts at the attached hydrocarbon chain, proximate to the aromat nucleus by a radical attack. This position is always very sensitive in similar structure, due to the fact, that the intermediate carbon radical is stabilized by the aromat and thus starts to stay persistent. As a fact, aromats may strongly boost oxidation of hydrocarbons if present in the mixture due to the mentioned persistency of the reactive intermediates. (Figure 17)

Aromats and naphtenics (containing unsaturated hydrocarbons and aromats) should be stabilized against oxidation.

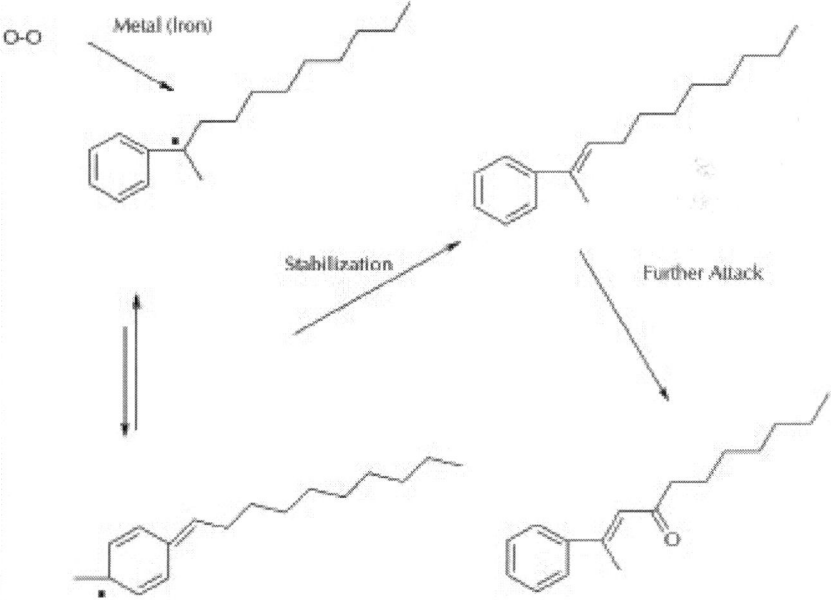

Figure 17: Oxygen attack in the oxidation mechanism of Alkylaromats.

SYNTHETIC HYDROCARBONS

Poly-Α-Olefins (PAO)

PAO is dominating all synthetic hydrocarbons by amount of production and worldwide turnover. Syntheses start from Dec-1-ene, a linear C10 hydrocarbon with a double bond at the beginning of the molecule. Polymerization and hydrogenation leads to PAO, as a highly branched and fully saturated hydrocarbon (Figure 18). [2, 4]

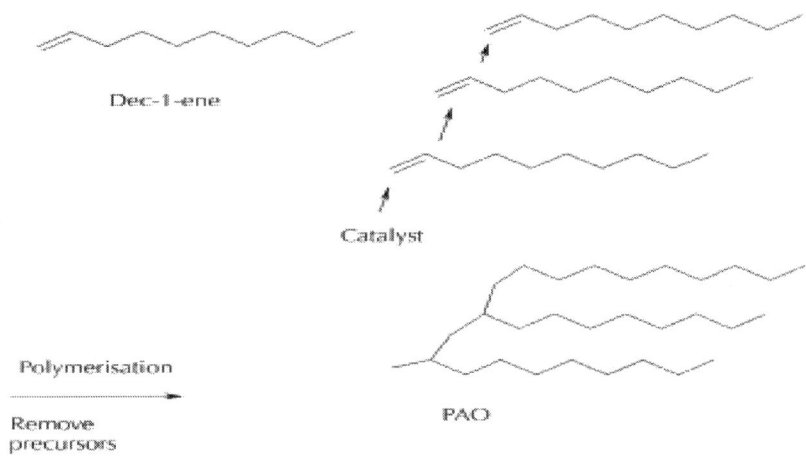

Figure 18: Principle of PAO formation.

Modern PAO may also start from a variety of hydrocarbons (C8-C12) by the same processes. PAO are the most prominent worldwide used hydrocarbons and found in all important applications, e.g. gear oils, circuit oil, hydraulic oil, base stock for automotive applications and others. [2, 3, 6, 7]

The extraordinary importance of PAO is due to its applicability at very low temperatures (pour points below –30°C) and, in the case of suitable antioxidant prevention also at higher temperatures (> 120°C). While PAO is, by its structure, very common in low temperature applications, it is very poor in the contact with metal surfaces beyond 120°C if not properly additivated by antioxidants.

Principal antioxidants for PAO are Phenyl-α-Naphtylamine (PAN) and octyldiphenylamines (see antioxidants (AO)).

Polyisobutenes (PIB)

PIB are a sub class of polymerized olefins. They are widely used to boost low viscous oils to higher ISO VG grades or as functional additives to improve the viscosity index (VI): the attitude of the lubricant to lower its viscosity strongly by temperature is reduced by addition of PIB. Synthesis is carried out starting from isobutene by catalytic oxidation processes (Figure 19):

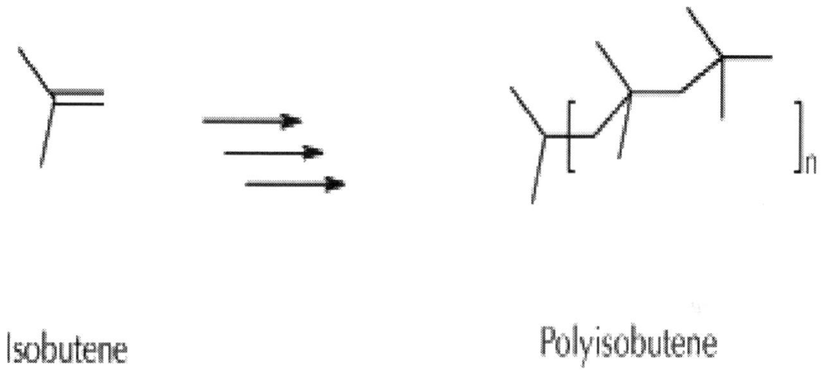

Isobutene   Polyisobutene

Figure 19: PIB formation by catalytic polymerization of Isobutene.

Sulfurization with activated sulfur precursors lead toward sulfurized isobutenes (SIB) widely used as extreme pressure (see also section about EP/AW additives).

ESTER OILS

General

Esters are in general reaction products between alcohols and acids. Their formation is also possible by means of other techniques, e.g. specific oxidation reactions, rearrangements in organic molecules or different reactions. [10]

Carboxylic Acid esters are created by the reaction of alcohols and carboxylic acids [A] and their derivatives, by trans–esterification (B), or catalytic reactions, e.g. epoxides with carbon dioxide (C) (Figure 20). [10]

A

Carboxylic Acid (A)   Alcohol (B)

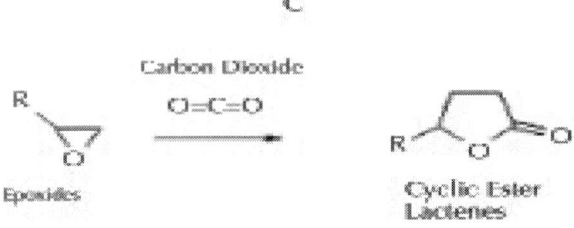

Formation of Carboxylic Acid Ester (C) by Reaction of Acid (A) and Alcohol (B)

B

Carboxylic Acid Ester (A)   Alcohol (B)   Carboxylic Acid Ester (C)

Formation of Carboxylic Acid Ester by Transesterification

C

Carbon Dioxide

O=C=O

Epoxides   Cyclic Ester
                  Lactones

Formation of Carboxylic Acid Esters by reaction of epoxides toward Lactones
(Cyclic Esters)

Figure 20:-Examples for creation of esters. (A) Reaction of Carboxylic Acids
with Alcohols, (B) Transesterfication, (C) Reaction of Expoxide to cyclic Esters.

Esters in Lubrication Technology

Despite their high variety in structure esters are used in different categories: [1, 2, 4]

Mono-Esters

Mono-Esters derive from a monocarboxylic acid (Carboxylic Acid that contains only one acidic centre) and monofunctional alcohols (Alcohole with only one OH group). [10]

Esters derived from this structure are seldom used as pure lubricants, more as solvents or dispersants. For example alcohol ethoxylates, formed by addition of alcohols to epoxides may be esterified by a monocarboxylic acid leading toward a dispersant or self-emulsifying solvent. (Figure 21)

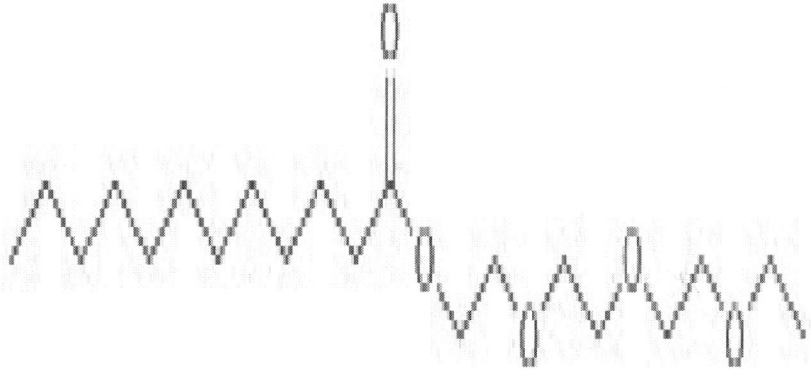

Figure 21: Mono Ester Formation with the specialty of esterified alcohole ethoxilates.

Di-Esters

Di-Esters are synthesized by use of dicarboxylic acids, mainly adipaic or sebacaic acid and two molecules of an alcohole. 2-Ethylhexylalcohole (Iso Octanole), leading to Di-isooctyladipate (DOA) or Di-isooctylsebacate (DOS, DEHS = Di-ethylhexylsebacate). (Figure 22)

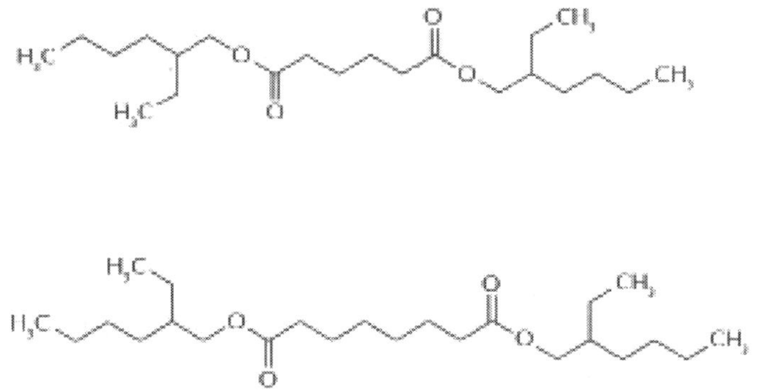

Figure 22: DOA and DOS.

They constitute an important group of oils, with either the function of base oil by themselves but also as adjuvant to mineral oil or PAO formulations.

For Di-Esters the reaction of alcohols (A) with two hydroxyl groups and a monofunctional carboxylic acid (B, B') is also applicable. (Figure 23)

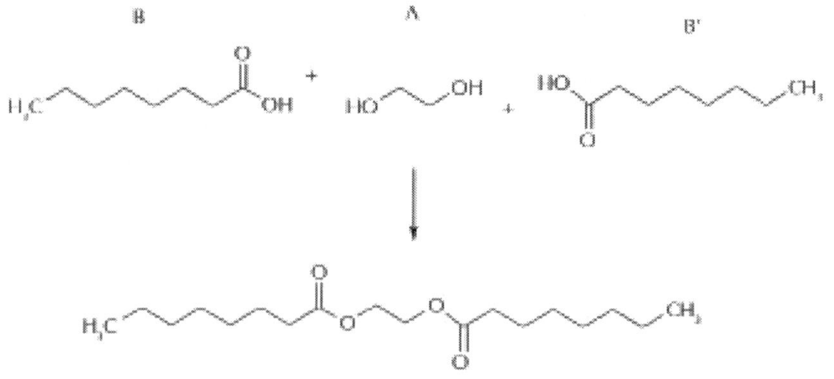

Figure 23: Formation of Di-Esters by Di-Alcoholes (A) reacted with Monocarboxylic Acids (B, B').

For technical purposes the reaction product of Neo Pentylgylcole (3.3 Dimethyl-propane-1.4-diol) with oleic acid is important in lubrication technologies for use as a friction reducer and in minimal lubrication systems. (Figure 24)

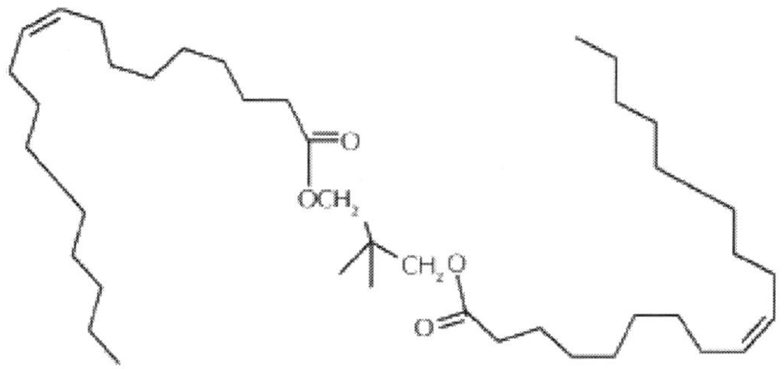

Figure 24: Neopentylgylcoledioleate (NPG-Dioleate).

Tri-Esters

Tri-Esters are mainly created by the reaction of trivalent alcohols with monocarboxylic acids. They are mainly represented by to major groups:

Glycerol Esters

Esters derived from glycerol as a trivalent alcohole leads to tri-Esters. (Figure 25)

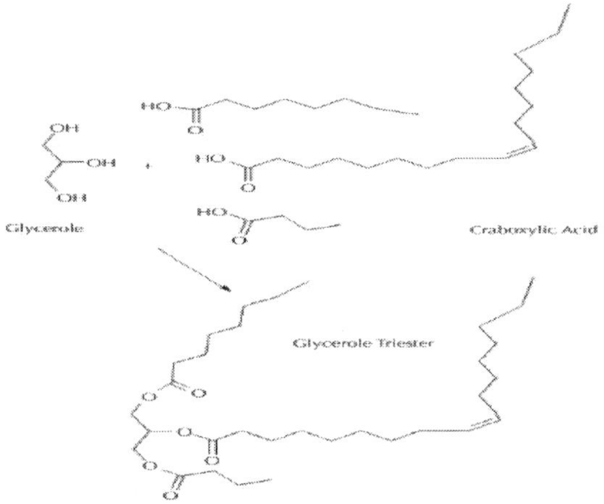

Figure 25: Esterification of Glycerole to tri-Esters.

Glycerol Tri-Esters represent the huge group of natural oils. Sunflower, rapseed oil are prominent representatives. A mixture of short chain carboxylic acids with unsaturated long chain acids is used.

As a fact of the presence of short chain carboxylic acids those esters are nutrients, biological degradable and widely used as natural, biodegradable oils.

As a special glycerol ester, important for lubrication, ricinoleic acid esters have to be mentioned.

Within this group ricinoleic acid represents the group of 12-hydroxy substituted C18 carboxylic acids.

Hence, alkaline cleavage of ricinoleic acid glycerol esters lead to 12-Hydroxi-oleic acid on the one hand and to sebacaic acid on the other hand by degradation of the double bond.

Catalytic hydrogenation of 12-Hydroxioleic acid results in the formation of 12-Hydroxistearic acid, which is important for modern grease concepts. Sebacaic Acid on the other hand is a raw material for DOS (see above) but also for the production of complex greases. (Figure 26) [1, 2, 4]

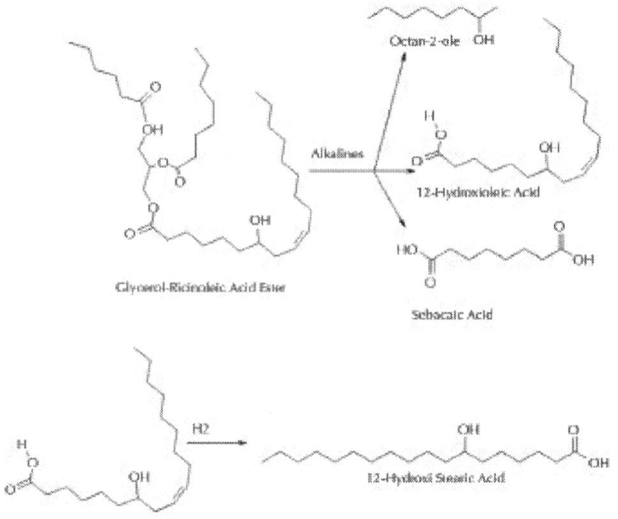

Figure 26: Cleavage of Glycerol – Ricinioleic Acid and hydrogenation to 12-Hydroxistearic Acid, Sebacaic Acid and Octan-2-ole [10].

Glycerole Esters with long chain carboxylic acids only, e.g. Glycerole Tristearate, are no longer nutrients and sparingly biodegradable. They are used as emulsifiers, consistency givers.

Glycerole Trioleate is a powerful friction reducer in tribological applications.

Triesters, Derived From Alcohols Else Than Glycerole

Trimethylolpropane Esters (TMP-Esters)

TMP-Esters are created out of Trimethylolpropane (TMP) by reaction with short chain carboxylic acids, e.g. the range from C6 to C10. (Figure 27)

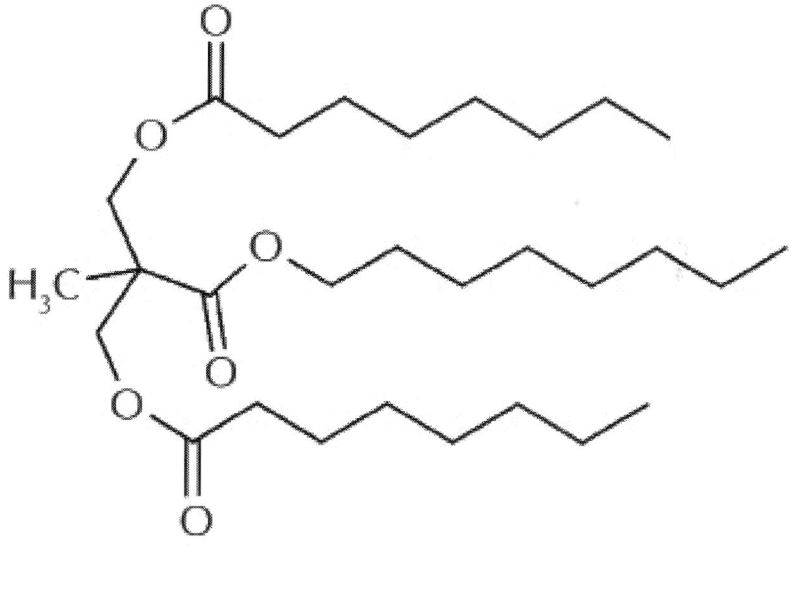

Figure 27: TMP Esters.

TMP Trioleate is created by reaction of TMP with oleic acid or by trans-esterification.and commonly used as lubricant in minimal lubrication.

Trimellitic Esters (TM-Esters)

Apart from the described structures where trivalent alcohols get reacted with monocarboxylic acids, trimellitic Esters (TM-Esters) are products from Trimellitic Acid Anhydride with Mono alcohols. (Figure 28)

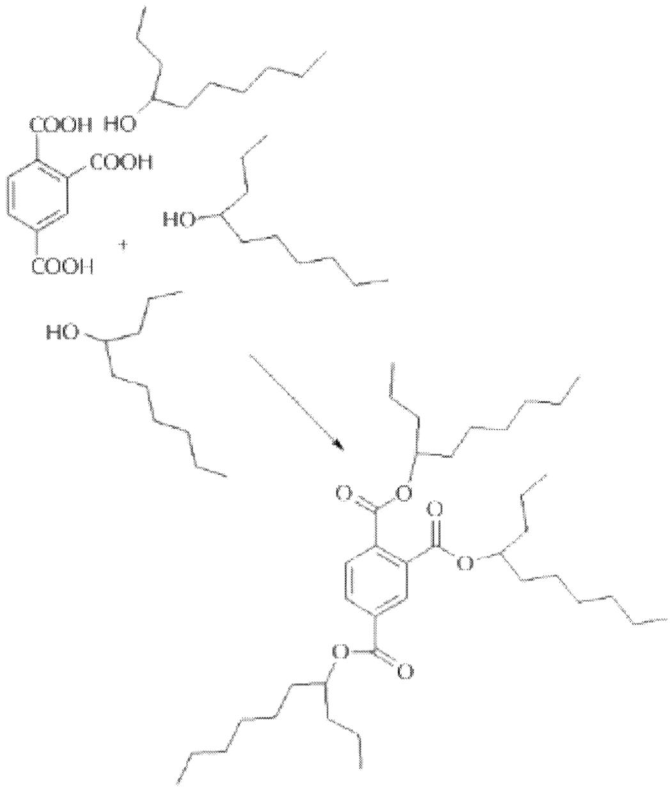

Figure 28: TM Esters by reaction of trimellitic acid with branched alcohole.

Due to the aromatic core those esters are high in thermal stability and widely used in high temperature applications.

Tetra Esters (Pentaerythrolesters, PE-Esters)

Pentaerythrole acts as a four-valent alcohole which may be esterified by four carboxylic acids. (Figure 29)

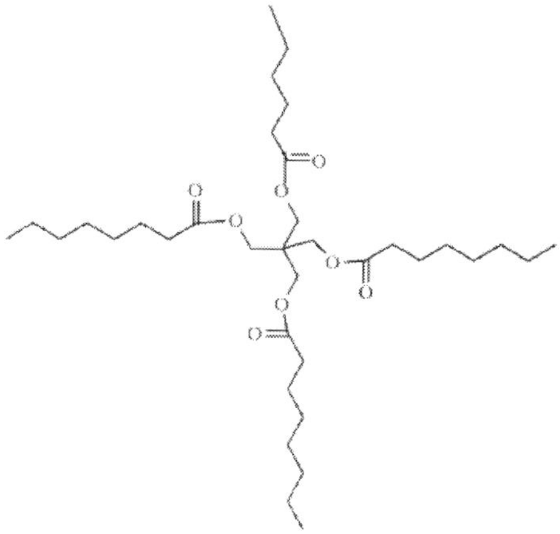

Figure 29: PE Esters.

Carboxylic acids are in the range from C6 to C10.

Dipentaerythrol Esters (Di PE Esters) are formed starting from Dipentaerythrole as a six-valent alcohole reacted by six monocarboxylic acids in a Carbon Chain length from 6 to 10. (Figure 30)

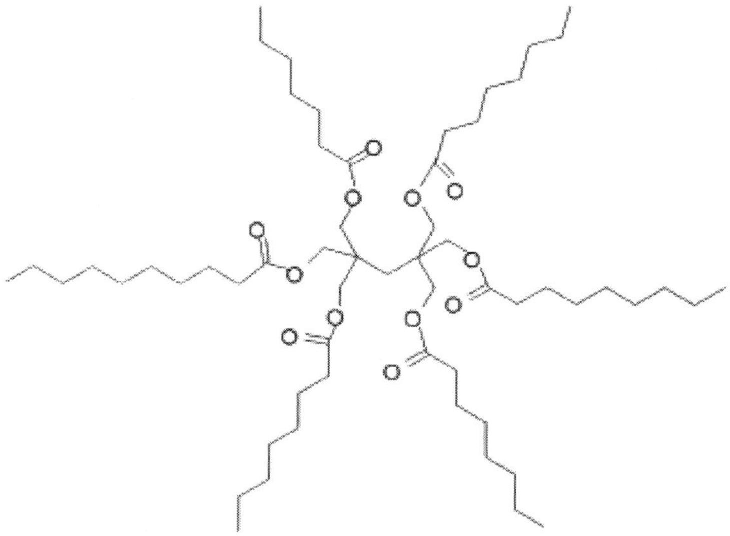

Figure 30: Di PE Esters.

Polyesters

In the past 20 years new groups of esters have been created by reaction of polycarboxylic polymers with alcohols. Those are reaction products of maleic acid anhydride (MSA) by Ene-Reaction with PAO precursors, leading to the PAO-backbone MSA addition product that might be esterified by butanole, leading to carboxylic complex esters (Figure 31).

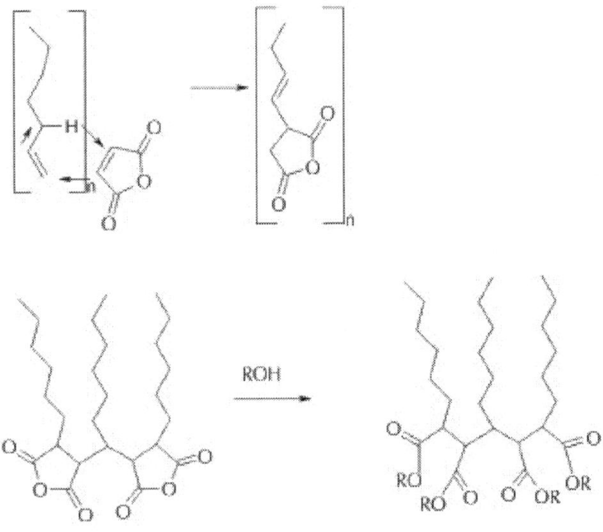

Figure 31: Complex Ester Formation by Ene-Reaction Sequences.

Complex Esters from those structures are widely used to improve the additive solubility and performance. Their structure with shielding the carboxylic groups causes less aggressiveness toward sealings.

STRUCTURE ACTIVITY RELATIONSHIP IN ESTERS

Esters are prominent representatives of lubricants where the chemical structure promptly leads to a specific tribological activity. However, if a tribological acitivity is demanded, the specific construction of esters may offer the solution.

Polar Acitivity

Esters are polar by their nature due to the central element where a carboxylic acid tail binds toward an alcohol. Polarity gives some advantage but also disadvantage in the case esters are used. In general, esters enhance the solubility of functional additives and keep them away from fall-out. Esters also enhance the cleaning of metal surfaces in operation, preventing a formulation by creation of sludge. Esters are, as a fact of their polarity, aggressive toward sealings with a general tendency to shrink them. Plastics and elastomers under bending are susceptible toward stress corrosion cracking if attacked by esters. Hence, stress-corrosion cracking has to be considered explicitly in the case if esters are used. Since hydrocarbons, like PAO have a tendency to swell elastomers, the addition of esters may counteract such that the effect is neutralized. As a fact synthetic oils based on PAO are additivated by addition of 10 or 20 % esters per weight to create this effect.

Low Temperature (Pour Point) Properties

Di-Esters, e.g. DOA, DOS are very useful in temperature ranges that undergo -40°C. This effect might be explained by the lack of hydroxyl groups that might associate at low temperature via hydrogen bridging, but also as a consequence of the crystallization hindrance due to the spatial structure of esters which does not allow a dense crystal packing. In contrast, esters may be designed such that their low temperature properties are lost, just by changing their structure. Also, if the number of polar groups increase the tendency to molecular association increases, and hence the pour point rises.

High Temperature Properties

High temperature applications in the use of esters are achieved by
- Sterical hindrance of the ß-Position in the Alcohol
- Use of Aromatic Nuclei in the Ester structure

As a specialty esters may rearrange within their structure via a preferred six-membered cyclic intermediate that creates an alkene on one side and a carboxylic acid on the other side (Figure 32).

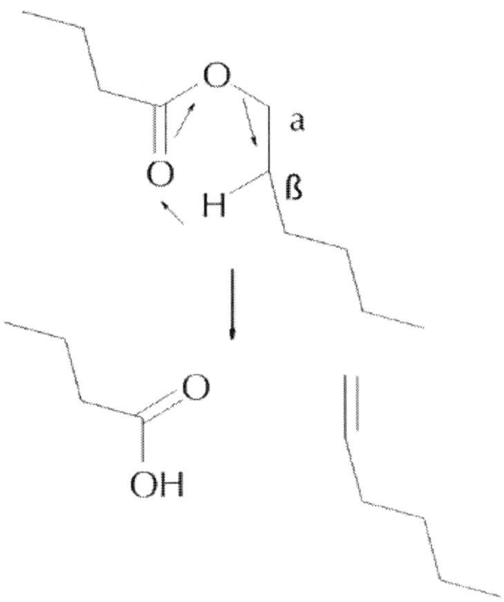

Figure 32: Decomposition of esters via cyclic rearrangement.

Degradation of esters via such mechanisms takes place at ambient temperatures, e.g. by copper activation even at 70°C. The formation of carboxylic acids and alkenes may lead to corrosion and unfavorable deposits on metals. In the case of blocking the ß-position, as in the NPG and TMP esters, the cyclic rearrangement is blocked and the ester does not undergo the thermal degradation. Such oils are commonly used as turbine oils.

Side Reactions

Hydrolysis

Ester Oils generally hydrolyze by interaction with water. The hydrolytic process is somehow the reverse reaction how esters form. The attack of water is enhanced if alkalinity is present but also acids may catalyze the hydrolysis. Common understanding states the attack of so called nucleophiles, like water at the carbonyl C-atom, followed by rearrangement sequences, leading to carboxylic acid and alcohols. (Figure 33)

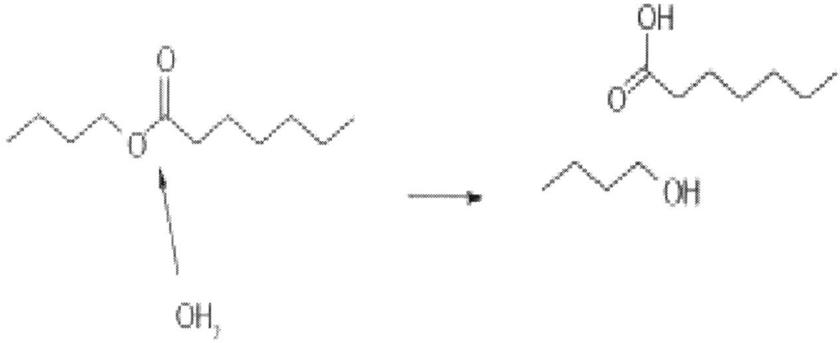

Figure 33: water-based cleavage of Esters toward carboxylic acids and alcohols.

Catalytic hydrolysis of ester oils also take place at metal surfaces, e.g. under tribological conditions. Formation of carboxylic acids may lead to corrosion as a consequence.

Biodegradation

Esters may decline under the interaction of bacteria and combust. Biodegradation is observed in the case of vegetable oils, e.g. glycerol esters, seldom on technical esters. In principal biodegradation cleaves esters, like water does to carboxylic acids. Biodegradation as a complex process does not stop there but lead to further products. Esters may oxidize as described in mineral oils and PAO at the organic tail. As a specialty they may undergo hydroxylation at a side position followed by trans-esterification to lactones. The lactone sequence is described already in the mineral oil section. Lactones are observed if esters, but also PAO are decomposed on iron at higher temperature. Infrared Spectra show absorption at $1800 - 1760 \text{ cm}^{-1}$ caused by lactone formation (see also chapter of antioxidants). (Figure 34)

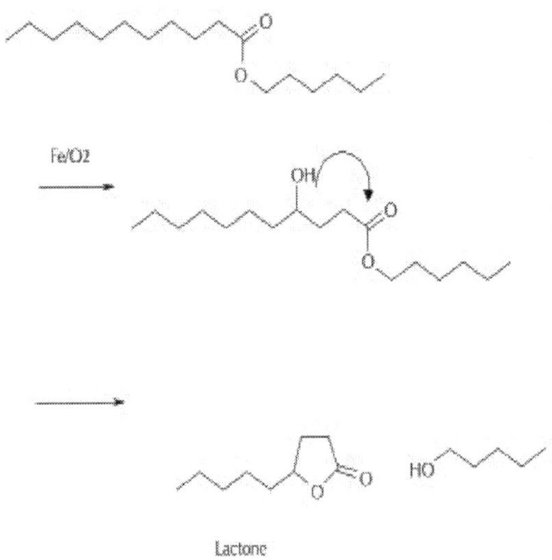

Fe/O2
⟶

Lactone

Figure 34: Lactone formation by side-chain oxidation of esters.

Other Esters

Esters may be created, as already mentioned by reaction of acids, in a different way as carboxylic ones. Prominent representatives are esters derived from phosphoric acid. Phosphorous offers two main oxidation states (+III and +V) from which acids are derived. Depending on the oxidation state and the alcohols, phosphoric esters are different in use. Also phosphorous overtakes the role of anti-wear activity in such substances. [9]

A common representative is Trilaurylphosphite (Figure 35).

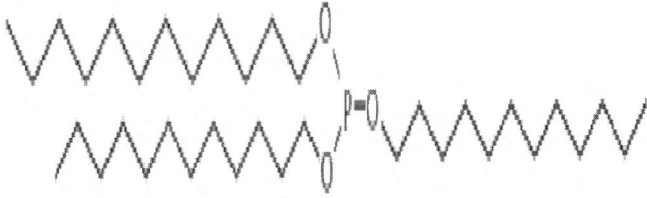

Figure 35: Trilaurylphosphite as a representative of Phosphinic Acid Esters.

Phosphoric Acid Esters

Phosphorus in the oxidation state (+V) creates a plenty of variant Acids, such as Orthophosphoric Acid, Diphosphoric Acid, Triphosphoric Acids switching into each other. (Figure 36)

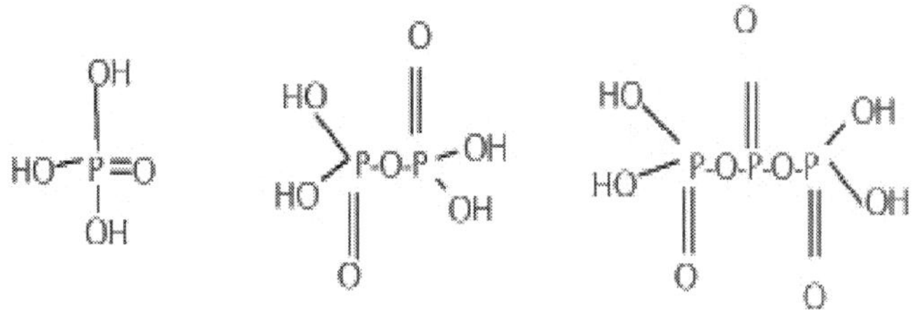

Figure 36: Representation of Phosphoric Acid Ester.

Phosphoric Acids are created by reaction of either phosphoric acid anhydride with alcohols or phosphoric acid derivatives, e.g. $POCl_3$ (Phosphorous-Oxi-Chloride) with alcohols. Aliphatic alcohols are in use, but also Phenols [10].

Formation of Phosphoric Acid Ester

Reaction of aliphatic alcohols, e.g. hexanole, with phosphorous pentoxide leads to hexylphosphate. In general some acidity remains due to insufficient esterifications. As a consequence those esters are often neutralized with amines to give amine phosphates. [10] (Figure 37)

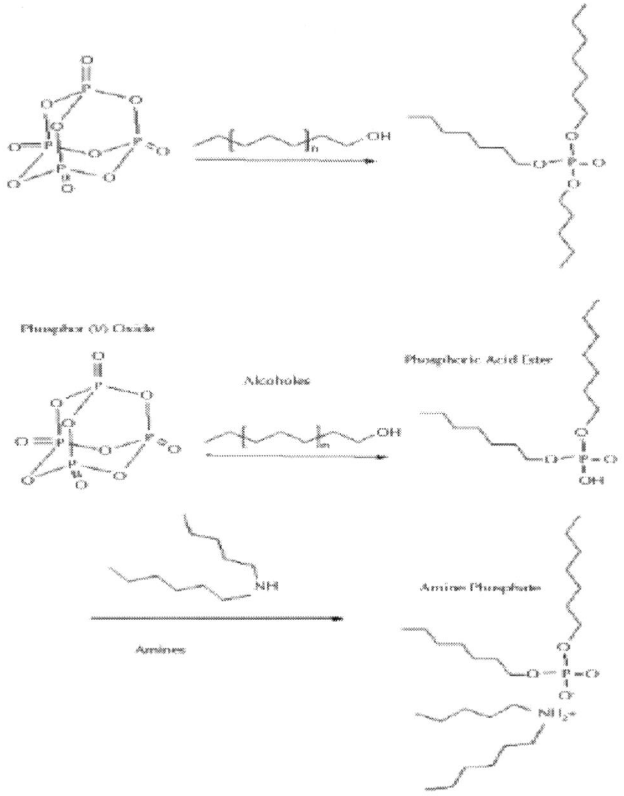

Figure 37: Phosphate Esters and Amine phosphates.

Amine phosphates are widely used as anti-wear and anti-corrosion additives in all kinds of applications. Phosphoric Acid Esters derived from Phenoles are shown below. (Figure 38).

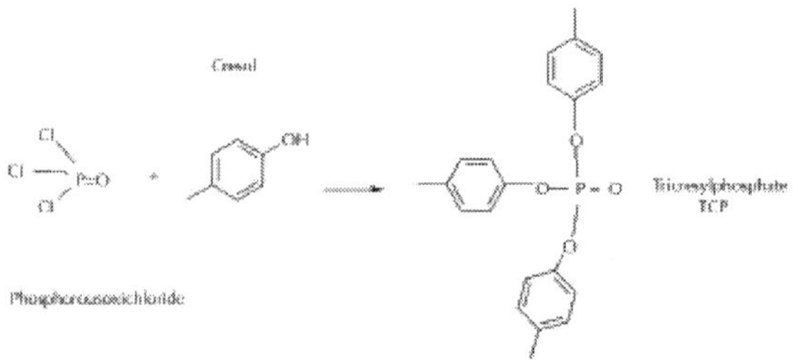

Figure 38: Arylphosphates derived from Phosphor Oxide Chloride Reactions.

In a different reaction Scheme Phosphoroxichloride reacts with alcohols. Those reactions are convenient to come to aryl phosphoric acid esters. Arylphosphates are somehow used to come to non-flammable high temperature lubricants at temperatures beyond 200°C.

Apart from the use as base oil phosphoric esters like Tricresylphosphates, based on the reaction from Phosphorous Oxichloride with Cresol (Methylphenols) are common additives for lubricants in bearing industry.

Whilst TCP with the methyl group in the para-position is seen as hazardous, TCP isomers in the ortho is registered to be highly toxic. Also mixtures of TCP isomers, due to the content of the highly toxic ortho isomer are registered as highly toxic. TCP, despite its superior behavior as AW additive for bearing lubrication is restricted for use (Figure 39).

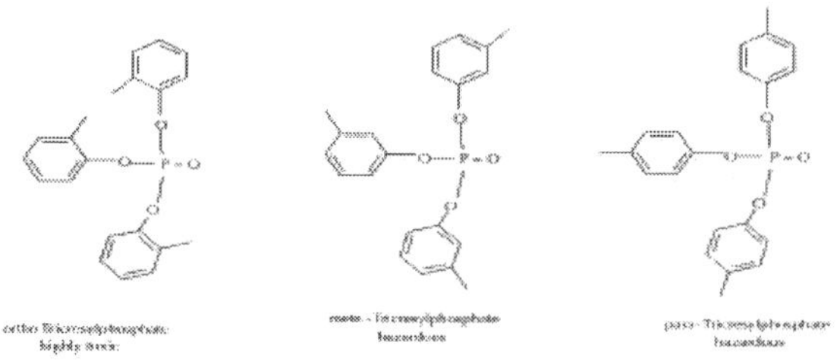

Figure 39: TCP and some isomers.

Use of Thiophosphorylchloride as precursor, the reaction with Phenole leads to EP/AW additives like Triphenylphosphorothionate (TPPT) and its derivatives. (Figure 40)

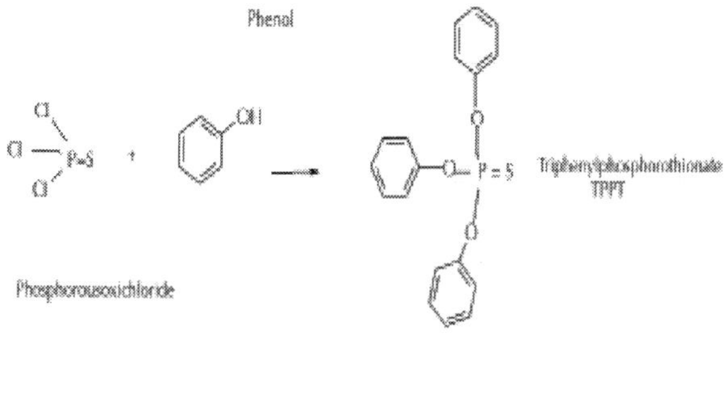

Figure 40: Synthesis of TPPT.

TPPT is widely used as non metal EP Additive as a substitute for Zn and Molybdenum Dithiophosphates. Due to its thermal stability, TPPT undergoes reactions at higher temperatures (>100°C). As to the fact that TPPT is ashless and starts to react at higher temperatures, it is a preferred additive in high temperature lubrication in combination with sterically hindered esters and PAO. In contrast to TCP, TPPT is not registered to be toxic, even more, the use of TPPT is allowed at level of 0.5% per weight for incidental food contact.

POLYGLYCOLES (PG)

General

Synthesis of Polyglycoles starts from Epoxides, obtained by catalytic oxidation of Alkenes from Petrol- or hydrocarbon chemistry. Polymerization catalysed either by acids or alkaline result in the formation of polyglycoles. In the case of alkaline catalyst, e.g. alkoxides on half of the PG contains a hydroxyl group while the end is capped by an ether function. (Figure 41) [1, 2, 7, 10]

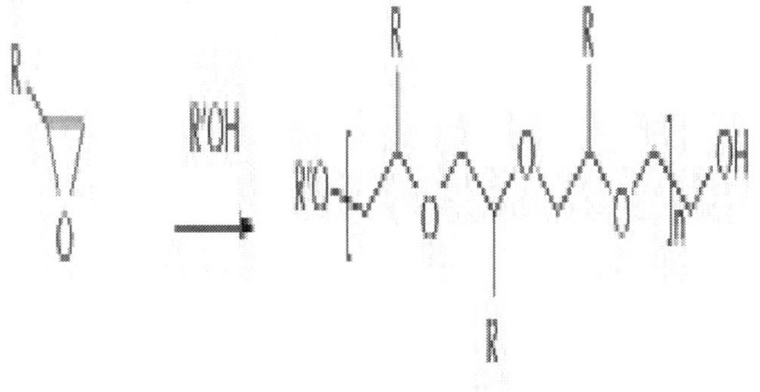

Figure 41: General Formation of Polyglycoles by alkaline catalytic polymerization of Alkene Epoxides.

The choice of either different alkenes (Group R) or alkoxides (R′) leads toward a huge variety of PG, all of them with different chemical and physical properties.

Table 1: PG and their data and applicability

Poitllltoles			PEG	PPG-PEG PEGPEGP136	PPG	PBG
Discriptian	Ethyleneciaide Polymer			Mixed Polymers Ethylene/ Propyleneoxide	Proyleneoxide Polymers	Butideneoxide Polymers
Chemical Data						
Physical Data	Density	aPPor-	1	0.95-098	095-0.98	095-038
	Flashpoint	approx.				
	PourpoiM	aPProx-				
	Wats miscible(704		100	partially		
	hydrocarbons		non miscible	partially	partially	partially
	Ests Oi ls		partially-full	partially-full	partially-full	partially-full
	Other PG	PPG	partially- full	partially-full	partially-hill	partially-full
		PBG	partially-full	partially-full	partially-full	partially-full

Tribological Data		32 - 46	32-100	32-100	32-100
Viscosity, 40C	Range				
VT-Coefficient	Range	180≥200	180≥200	180-->200	180-->KO
VP Coefficient	Range				
Seals	MBR	compatible			
	ABS	compatible			
Paintings		not compatible			
Others					

Technically only a couple of variances are produced in a larger scale, such as:

- *Polyethylenglycoles (PEG)* where Ethylene Oxide is the starter
- *Polypropyleneglycoles (PPG)* where Propylene Oxide is the starter
- *Polyethylene- Polypropylene Oxide Mixtures* started from mixtures or Ethylene and Propylene Oxide

Table 1 offers an overview across the most common PG their data and applicability.

Single addition of long chain alcohols lead to the formation of fatty alcohol ethoxilates, for use as non-ionic detergants and dispersants in lubricant formulations, as silicone free defoaming and emulsifiers for lubricant formulae.

In general PG are not thermally stable by themselves and tend to decompose by emission of volatile degradation products, e.g. low boiling compounds, such as aldehydes, ketones, acids and others. Due to this behavior PG are used in high temperature applications where the formation of polymers and lacquers due to heat induced degradation of lubricants is not convenient, for example high temperature chain lubrication.

Presence of alkalines, such as overbased sulphonates, widely used in motor oils, as corrosion inhibitor lead to multiple cross-reactions with the decomposition products of PG (aldol reactions): Results of the aldole reaction are tars, sludge and slurries in the system. In consequence corrosion resistance of PG should always be carried out by acidic corrosion inhibitors, such as succinic-esters, Zinc-Naphtenates or Phosphoric partial esters. (Figure 42)

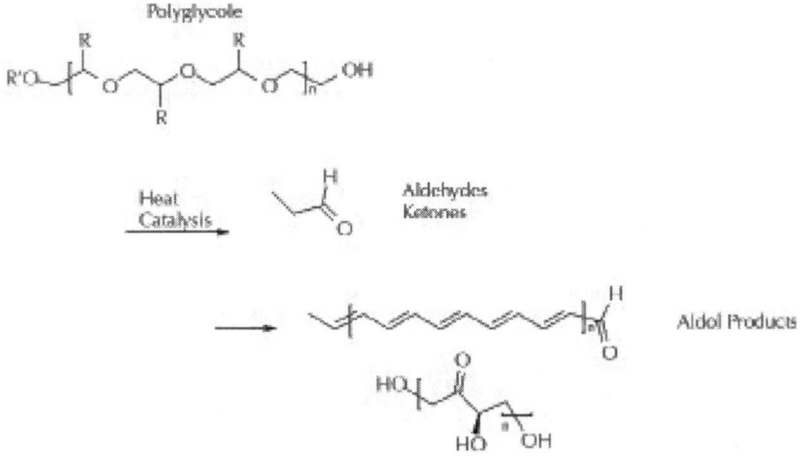

Figure 42: Aldole sludge formation in PG by use of alkaline.

It has to be considered that PG are poorly soluble even amongst themselves and should be carefully checked. In general their solubility in mineral oils is poor, better in esters (depending on the structure). However, PG needs to be stabilized by antioxidants in order to prevent the early thermal degradation. By doing so, the application of PG are enhanced significantly, such, that even applications temperatures > 160°C are approached.

Convenient stabilizers are Phenyl-α-Naphtylamine, Phenothiazines or Alkyldiphenylamines. The amount should be adapted to the application.

In general PG offer very high viscosity indices, mainly above 160 (compared to mineraloils at ranges from 20 (alkylnaphtalenes), napthenics (70), paraffine base solvates (110), Poly-α-Olefines (140).

This high VI allows reducing the calculated viscosity in a given tribological application down to one or two levels. For example, if in a given application ISO VG 320 (320 mm²/s, at 40°C) is calculated for a mineral oil with a VI of 100, this viscosity maybe reduced by use of a PG down to 220 mm²/s or even 150 mm²/s. Pour points are low in the case of PG, very often in the range of -30.. – 40 °C even. Reaching the Pour point, PG tend to form highly viscous liquid, however, crystallization-inhibited. As a fact of this huge increase, PG is not for use even at temperatures above the pour point. Realistically

PG is not suitable in the vicinity of their pour point. Therefore they are not very good low temperature base oils compared, for instance, with esters or PAO fluids. Due to their chemical structure PG are somehow strong solvents, e.g. paintings. In the case PG is used, the system has to be checked whether the paintings of the tank, the machine housing or others are affected. Dissolved painting from the tank may cause severe problems in the oil circuit by blocking filters. Additive response, known from standard applications, may change seriously by use of PG due to their different solvent capability. Extreme Pressure Additives have to be checked in their performance if used in PG. Normally anti-wear and anti-friction additives may be decreased in their content.

As a fact of the presence of epoxides in PG and due to their cancerogenic potential, the use of PG formulations drops down.

Polyethylene Glycols (PEG)

Polyethylene Glycols are made from ethylene oxide by polymerization (Figure 43).

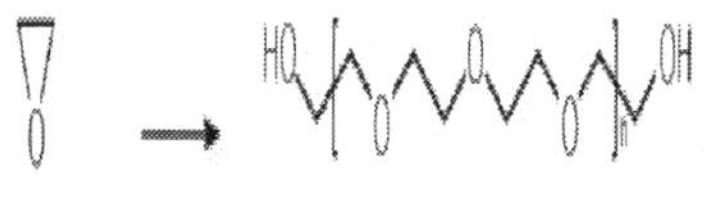

Figure 43: PEG formation and structure.

PEG, apart from its wide use in cosmetic industry is completely water miscible. Due to its water miscibility PEG is only or sparingly soluble in hydrocarbons. Compatibility of the PEG with a given fluid has to be checked before use. PEG, as facts of its water miscibility will uptake water without separation. In case of the use of PEG in applications within water environment the water ingress should be checked carefully. Effects of water ingress are increasing threat of corrosion and thinning due to the mixture.

Use of PEG

Water miscibility is of use in non-flammable hydraulics in coal mining industries, but also in applications of pharmacy and food processing.

In general PEG is allowed within the FDA regulation to be safe for incidental food contact. Due to the positive effect of sliding especially in worm gears PEG is somehow recommended for use in such applications. [2, 6, 7, 9]

Polypropylene Glycoles (PPG)

Polypropylene Glycols are made from Proplyene Oxide by polymerization by use of butoxides leading to a half ether structure (Figure 44):

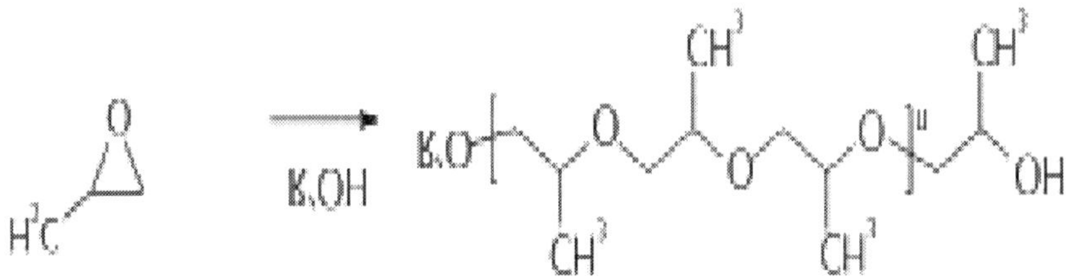

Figure 44: PPG Structure.

Due to the additional methyl group in the structure water miscibility drops down (contrast to PEG) while the oil miscibility promotes. Also by choosing longer alkyl chain butoxides, PPG structures may be obtained with enhanced oil solubility.

While PEG is highly dissolved in water, PPG forms droplets immersed in the water. Due to this fact water separation out of PPG is difficult to achieve.

The partial solubility and immersion of PPG in water causes a very high fish toxicity. PPG should never be used in the case of its break out-in free lands or water

Use of PPG

PPG is commonly used as high temperature circuit oil, e.g. calandars, compressors, high temperature chain lubrication. All over PPG has to be stabilized by acid corrosion inhibitors, e.g. phosphoric partial esters and antioxidants like Phenyl- -Naphtylamine [2] [3] [5] [8].

Polybutylene Glycoles (PBG)

PBG are seldom in use and made consequentially from Butylene Oxide polymerization by use of alkoxides, leading to half esters (Figure 45)

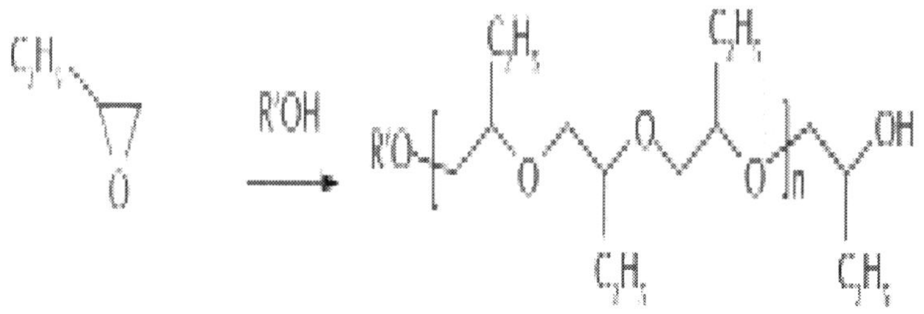

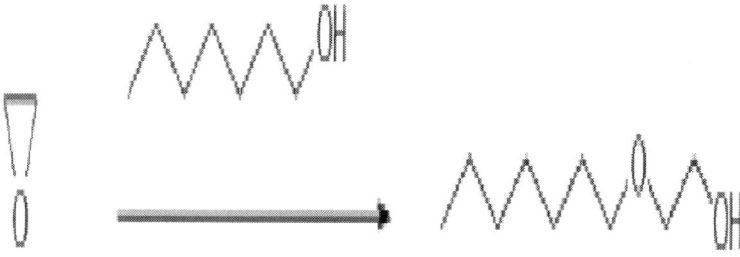

Figure 45: PBG Structure.

Use of PBG

PBG is useful for enhancing the solubility of additives, boosting the viscosity index of mineral oil variants.

Alcohole Ethoxilates

Alcohole Ethoxilates are formed by a cross reaction of epoxides with alcohols. (Figure 46)

Figure 46: Alcohole Ethoxilates.

The use of long chain alcohols leads to alcohole ethoxilates being used as non-ionic surfactants, emulsifiers and dispersants in multiple applications, e.g. hydraulic oils with dispersant capability, cutting fluids, and dispersants for applications where sludge is expected.

Due to their non ionic nature alcohole ethoxilates are widely compatible in lubricant formulations.

SILOXANES

Siloxanes (Silicone Oils) are common lubricants in multiple applications, such as food and pharmacy, but also in applications where special low friction properties are demanded [2, 3, 8]

In general silicones are the result of alkylchlorosilane hydrolyses [10] (Figure 47).

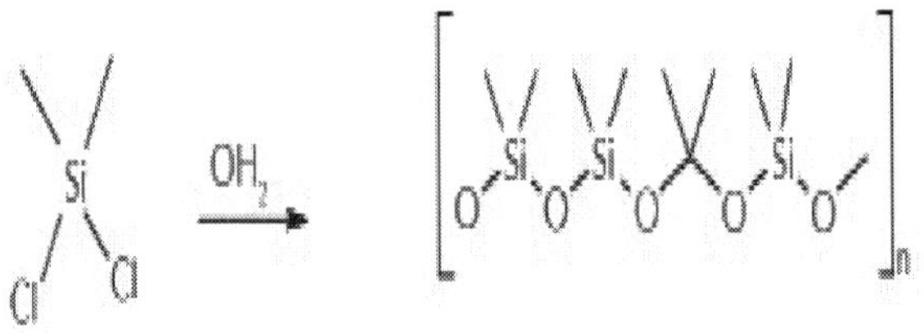

Figure 47: Scheme of Silicone Oil formation.

Side groups are methyl, but also phenyl groups leading to polydimethylsiloxanes or polyarylsiloxanes. Mixtures of methyl and arylsiloxanes are in use with different spreading between in the side chain (Figure 48).

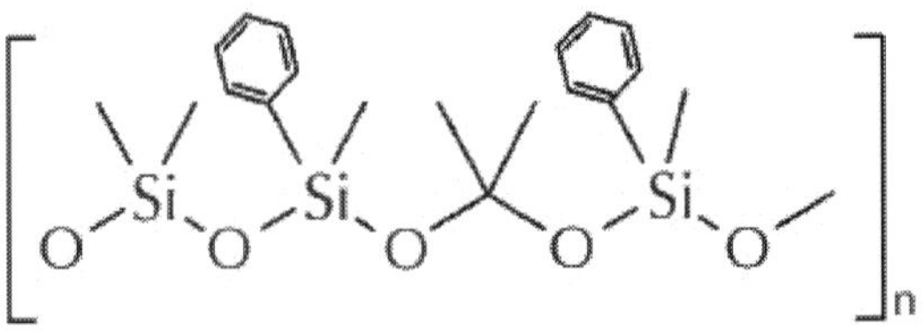

Figure 48: Polymethyl-Aryl Siloxanes.

Siloxanes with different structures are not generally miscible amongst each other. Miscibility has to be checked carefully. As mentioned, siloxanes are widely used in lubrication technology due their exceptional properties concerning low friction capability, high temperature stability and low toxicity in various applications. Prominent applications are starter components in cars, valves in food industry, slow speed bearings and high temperature applications where arylsiloxanes are in use. Siloxanes creep widely across surfaces and may cause problems in coatings, lacquering and paintings.

POLYFLUORINATED POLYETHER (PFPE) BASE OIL

PFPE Base Oil is created by polymerization of Perfluoroepoxids. Structure of PFPE is similar to polyglycoles but with overall substitution of hydrogen by fluorine [2] [3] (Figure 49)

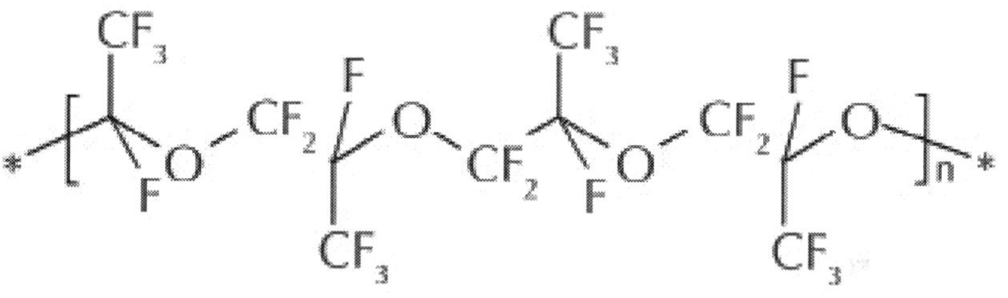

Figure 49: PFPE Base Oil.

Due to the effective shielding of the C-O-C backbone in the structure of PFPE by the trifluoromethyl side chain group PFPE are completely insoluble in water, inert toward alkaline and acids and even oxygen.

PFPE Base oil is used for high temperature purposes and in the presence of aggressive media, mentioned above in junction with PTFE thickener. (Figure 50)

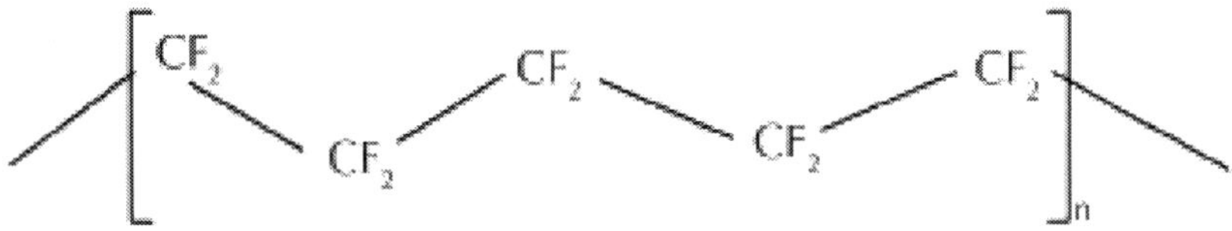

Figure 50: PTFE as thickener for PFPE..

PFPE sparingly adheres to metal surfaces due to droplet formation. The low adhesion causes creeping across surfaces and mal-lubrication if the surfaces are not cleaned thoroughly. Creeping and low adhesion may cause low friction in certain applications. PFPE is insoluble in most of the common base oils. Use of PFPE hence is restricted to the fluorine group of base oil.

Inertness of PFPE and PTFE make greases suitable for incidental food contact lubrication.

High temperature combustion of PFPE may cause the emission of hydrogen fluoride and fluoro phosgene which makes PFPE formulations somehow corrosive, especially on steel alloy compositions. Due to this fact, the high temperature corrosiveness should be carefully taken into account in the case of PFPE use.

ADDITIVES

Additives in lubricants enhance base oil functionalities. Additive technology is in broad scale based on organic chemistry syntheses. From their origin they are found by chance, less than by a real scientific approach. Nevertheless, literature about their reactions is innumerous from the very beginning. [2, 3,9]

Modern additive technology commenced in the early 20th century and has progressed continuously due to advanced organic chemistry syntheses. Upcoming modern spectrometry has been used to clarify their structures and their reaction at different metal sites. [2, 3, 9]

Beyond the basic and industrial reaction mechanism studies, mixtures of additives have been studied extensively by industry and science over the years. Such studies reveal the mechanisms of compatibility and incompatibility of additives acting together at a given application. [9]

For example a functional mismatch is caused by diverse demanding addressed to additives, e.g.: additives acting against corrosion may interfere with additives that have to prevent metal surfaces against fretting or welding.

Modern additive technology is inevitable to reach the "for-life" goal of modern technologies. As "for-life" might be understood in a different manner by users, additive packages are developed during the decades adapted to a given customer demanding. For example, the demanding to get automatic transmission gear oil performances is achieved by additive packages that may not fit for wind turbine or paper mill applications. Hence, additives and their mixtures have to be selected carefully for each purpose. [2, 3, 8, 9]

In general there are no rules up to now to predict additive performances at a given technical application. As a consequence formulations have to be tested in forecast extensively to assure its functionality. Such testing is addressed by international and national regulations.

Additives may cover a distinct structure-property relationship. Since there are no scientific rules declaring on how a chemical structure of an additive causes a function all variations in additives have to be validated by tests.

Additive technologies have been revised many times during their history, either due to a change in demanding or due to their toxicity. Toxicity is a severe problem in additive technology, since no one knows their real long term biological and ecological effects. [9]

Since the validation of those different chemical additive structures causes tremendous costs, it is a fact, that additive free technologies or additive technologies with marginal content level are favored as future solutions.

The following chapter addresses additive technologies concerning extreme pressure, anti-wear functions and also corrosion-protecting and antioxidants.

Extreme Pressure (EP) and Anti-Wear (AW) Additives

General

Extreme Pressure (EP) and Anti-Wear(AW) Additives are functional chemicals in lubricants with the task to separate metal surfaces in the case of heavy loading and to improve their resistance toward wear in the case of oil film break in the contact [9].

Machinery elements that start to run or stop due to emergency show pronounced loading due to a lack of lubrication, e.g. the oil does not separate the metal surfaces and the protection of the oil film drops down. At that point EP and AW additives are supposed to jump into the arena by causing reaction layers preventing the metal from direct rupture or welding.

Their chemical structures are found by chance. For example observations during drilling and maching show that tools perform better if lubrication is carried out by use of sulfurized oils derived from vegetables, mixed and heated with sulfur.

Later on intense research the nature and reaction started including modern surface spectrometry techniques. The transformation of EP/AW additives as a function of the nature of the surfaces, their loading, contact geometry, temperature and their structure shows a clear picture of structure-activity relationship. Also additives perform as a function of their chemical structure, but also as a function of their solubility in base oil and as a function of other additives being present. In that sense, it is shown that additives either may prolong service life but are also capable to shrink life.

Sulfur Additives

Sulfur acts as a powerful extreme pressure additive. The high reactivity, especially toward copper makes it unlike to use sulfur as element in tribology.

Sulfur embedded in organic framework acts as a powerful Extreme Pressure additive. Choosing appropriate organic structures the activity toward copper drops down. However, using sulfurized additives copper deactivation should be present anyhow.

Sulfur is added either by reaction of reactive organic precursors like alkenes and their derivatives by heating up with the element, or by polymerization sequences with activated sulfur precursors such as di-sulfur dichloride. Doing so, all kinds of unsaturated specie gives reaction products leading to sulfurized specie. Prominent representatives are reaction products of Isobutene with Disulfur Dichloride, or reaction products with terpenes (Figure 51) but also unsaturated carboxylic acid esters, like rapseed oil:

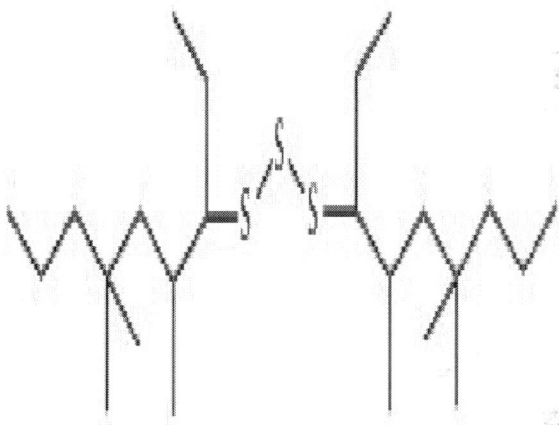

Figure 51: Didodecyltrisulfide as polysulfide representative.

Sulfurized Additives (S-Additives) are often used together with phosphoric acid esters, since the synergistic between those additives are known from the past. Doing so, gear oils may contain S-Additives with amine phosphate esters. Also extreme pressure additives containing Dialkyl-Thiophosphoricacidesters are prominent representatives in sulfur additive chemistry. (Figure 52).

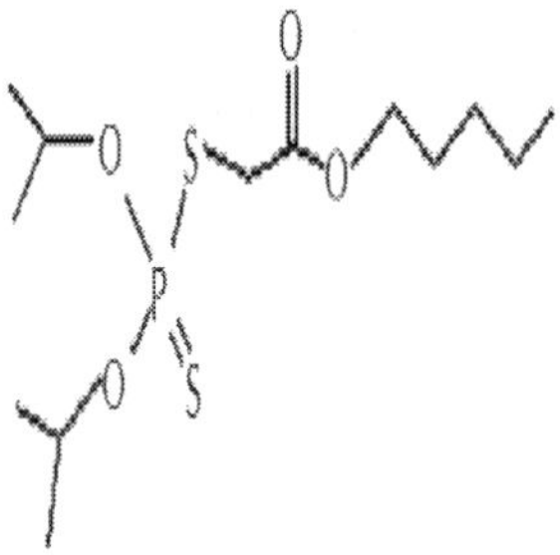

Figure 52:-Thiophosphoric Acid Ester.

Dithiophosphates

Zinc- And Molybdenum Dithio Phosphates (ZndtP- Modtp)

Zincdithiophosphate (ZndtP) represent a prominent group of EP/ AW additives. They derive from the neutralization of Thiophosphoric Acids, obtained by ring opening of Phosphorous pentasulfide with alcohols, with Zn-Carbonate or Hydroxides. As a fact, the ZndtP differ strongly by their carbon-chain length. A couple of variants are achieved by choosing different alcohols in the ring opening sequence of Phosphorous (V) sulfide. From the structural perspective, ZndtP may be regarded as chelate complexes rather than a salt (Figure 53).

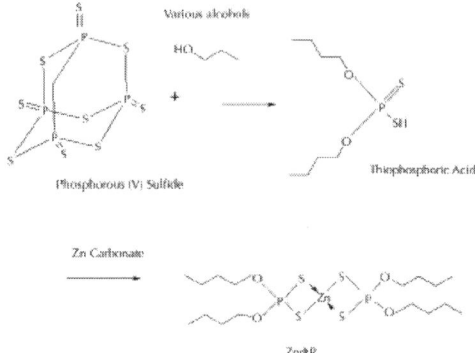

Figure 53: ZndtP from neutralization of Thiophosphoric Acid with Zn Carbonate.

Molybdenumdithiophosphates contains a Molybdenum [µ-oxo] Core, distinct compared to ZndtP (Figure 54).

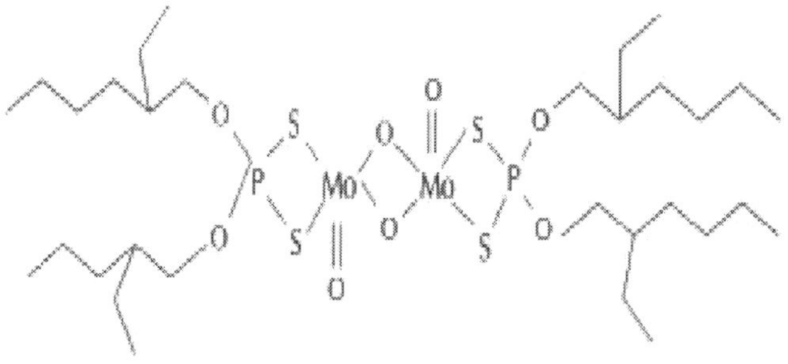

Figure 54: Molybdenumdithiophosphate.

Dithiocarbamates

Similar to Dithiophosphates, Chelat Complexes from Zinc, Molybdenum but also Bismuth and others may be formed by reaction of Thiocarbamic Acid with the metal precursors. Dithiocarbamic Acid is synthesized via addition of amines to Carbondisulfide. By varying the chain length of the amine different dithiocarbamates are achieved (Figure 55 and Figure 56):

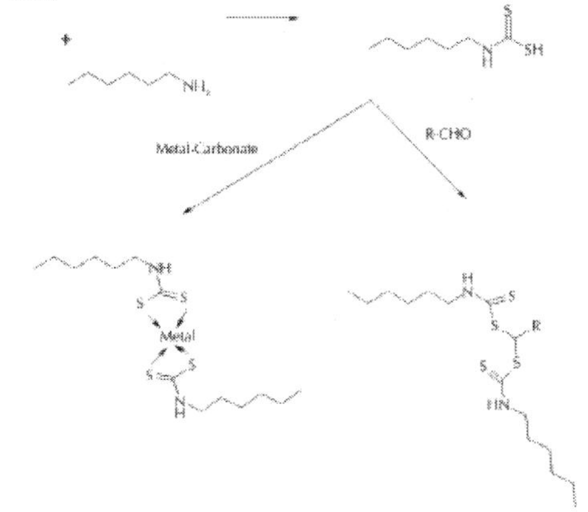

Figure 55: Syntheses of dithiocarbamates.

ZndtC

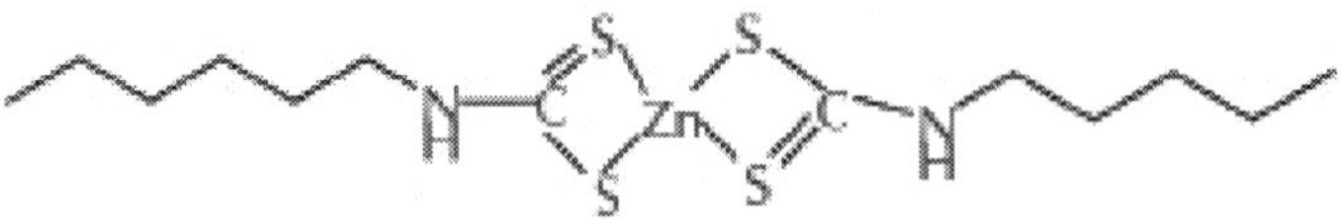

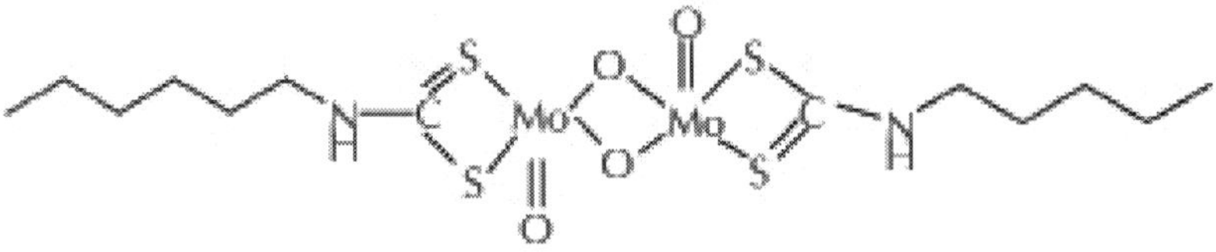

ModtC

Figure 56: Zincdithiocarbamate (ZndtC) and Molybdenumdithiocarbamate (ModtC).

Corrosion Protection

General

Within this chapter only iron as a chief element in technical application is considered.

Generally metal surfaces tend to corrosion if water, oxygen and probably salts, like sodium chloride are present. Corrosion may take place either by cathodic reduction of oxygen or by anodic oxidation of the metal. Charges, either positive (anode) or negative (cathode) pass the surface layer. [2, 3, 9]

Charge transport from the metal toward the outer region is hindered by the surface potential (over potential). Thus, corrosion processes have to overcome this potential and start after a certain induction period. Once, if this potential has been overcome the corrosion starts without hindrances by successive material transport. Materials transport ends up in a drastic change of the surface, mainly accompanied by a loss.

For iron as metal, the transport of the metal ends up in a flaky layer (Rust) that permits water and oxygen to penetrate. Due to this effect the rust process ends up in a total damage of the metal, especially in an environment that boosts corrosive processes.

Counteracting corrosion, the initial processes of charge transportation have to be blocked. Doing so, the over potential, e.g. the natural barrier of charges passing the surface has to be increased by creating additional layers on the metal surface (Passivation) or by creation of stable, insoluble complexes, formed by interaction of the surface atoms with a complex builder.

Passivation of iron surfaces and enhancing the over potential is achieved by deposition of chromium layers that cause a thin and gas-dense closed layer on the metal. Thus, chromium is a powerful inhibitor toward corrosion processes. As the charge transport phenomena occurring on iron surfaces are cathodic or anodic and vice versa, this process could also be stopped by offering an anodic victim like a zinc coating.

Additives that create a corrosion protection are in general dissolved in a carrier base-oil that spreads over the surface. Due to their adapted functional groups a physical binding toward the surface starts to create a layer. In order to create an appropriate corrosion protection this layer has to be packed dense to avoid the penetration of water and oxygen. This is realized by strong dipolar groups and oil soluble tails with a marginal demand in lateral spacing, e.g. long, - unbranched alkyl chains.

Else, passivation also is achieved by placing insoluble complex builders onto the iron surface, like phosphates are. Iron phosphate builds up a close dense insoluble layer on the surface.

Restriction of iron phosphate is indicated by the fact that, under certain conditions, phosphates start to get reduced forming posphanes. Phosph4anes strongly affect metals due to segregation of phosphorous at grain boundaries and releasing hydrogen into the metal. Hydrogen is detrimental to the microstructure by inducing, e.g. hydrogen enhanced local plasticit (HELP) or hydrogen induced cracking (HIC). Presence of phosphanes by reduction of phosphates takes place in acidic and reducing environment, e.g. presence of hydrogen sulfide, chlorides and others.

The following chapter will show some of the most prominent representatives of corrosion protectors.

Sulfonate-Chemistry

General

Sulfonates derive from sulfonic acids by neutralization with alkali, earth alkali –metals but also with metals from the transition group, for example zinc. Principally each sulfonic acid may be neutralized. In technical applications mainly alkyl benze sulfonic acids and dodecylsulfonic acid are neutralized. Production starts from alkenes out of petrol chemistry by addition sulfuric acid or SO_3.

Neutralization with either sodiumhydroxide, Calciumcarbonate, Magnesiumcarbonate or Bariumcarbonate leads to sulfonates: A = Sodiumsulfonate, B = Calciumsulfonate and with excess Carbonate to over based Calciumsulfonate (B′), Magnesiumsulfonate (C) and Bariumsulfonate (D) (Figure 57).

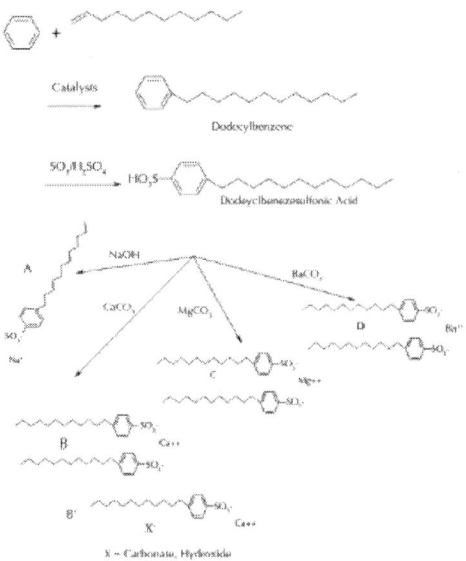

Figure 57: Sulfonic Acids and their Salts.

Carboxylic Acids and Derivatives

Carboxylic Acid and their derivatives, e.g. esters may act as metal corrosion protectors. While carboxylic acids are supposed to cause corrosion, some of them prevent. Rust preventing carboxylic acids are derivatives from α-Aminoacids, like N-Oleylglycine. (Figure 58)

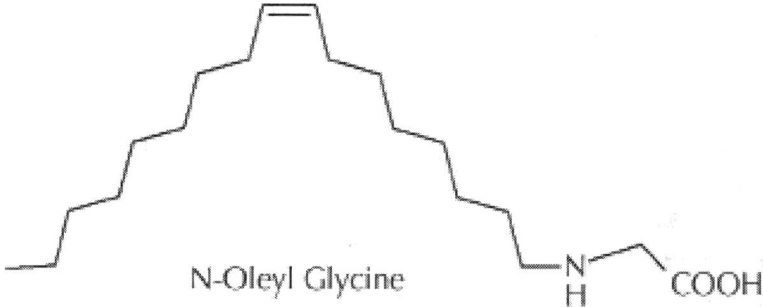

Figure 58: N-Oleylglycine.

N-Oleylgylcine acts as powerful emulsifier, even at low dosage. Rust protecting is due to the spread of water in the formulation over a big volume. N-Oleylglycine, even at low percentages also counteracts with EP/AW additives, driving their activity down.

Carboxylic Acids, derived from Phenoles such as Nonyl-phenoxiaceticacid is a non-emulsifying corrosion protector but under prohibition, due to its irritating effects (Figure 59).

Nonyl Phenoxy Acetic Acid

Figure 59: Nonylphenoxyaceticacid.

Succinic Acid Derivatives, such as Succinic Half Ester of Octanole are powerful metal protectors, but also strong counteracting with EP/AW additives. Synthesis is carried out by reacting succinic acid anhydride with alcoholes (Figure 60).

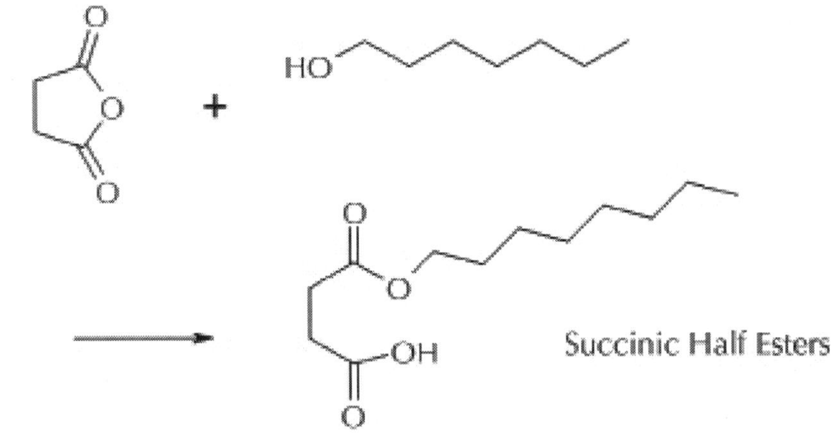

Figure 60: Succinic Half Ester.

Carboxylates, derived from neutralizing carboxylic acids with transition metals like Zinc, Lead, Bismuth lead to corrosion protection. Common acids are Napthenic acids or medium chain carboxylic acids like octanoic acid (Figure 61).

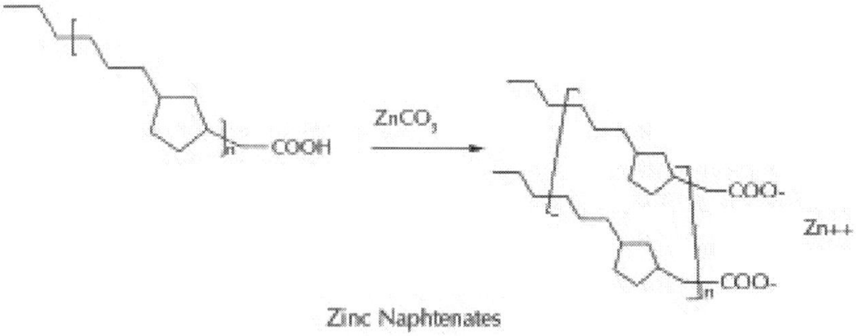

Figure 61: Zn (Bi) Carboxylates (Napthenate and Octoate).

Amine Phosphate Esters

Amine Phosphate Esters may act as anti-corrosion additives in addition to their anti-wear properties. Due to their synergistic properties and due to the fact, that certain amine phosphates are allowed as additives for incidental food contact, they are often found in all kind of lubricants (Figure 62).

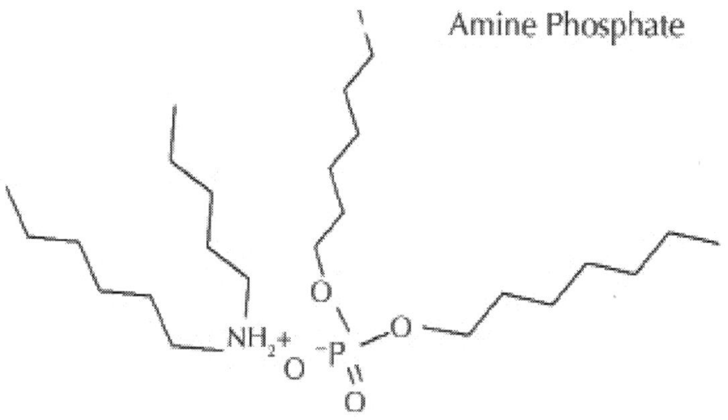

Amine Phosphate

Figure 62: Amine Phosphate Structure.

Amine Phosphates are powerful activators of copper and zinc and cause leaching of those metals from brass cages in bearings. Adding Amine Phosphates copper deactivators like benzotriazoles have to be present (Figure 63).

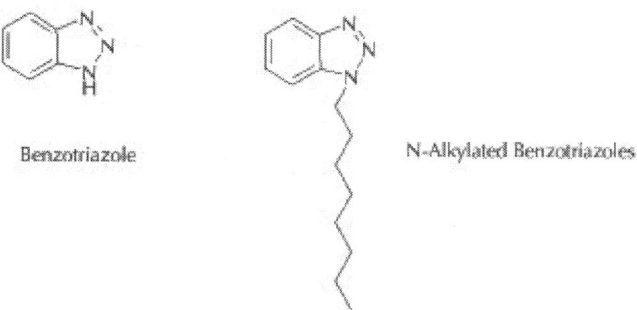

Benzotriazole   N-Alkylated Benzotriazoles

Figure 63: Benzotriazole and N-Alkylbenzotriazoles as Cooper Passivators.

Antioxidants (AO)

AO prevent lubricants from oxygen attack. Oxygen is, by nature, a diradical that undergoes several transitions. Electron uptake from metal surfaces by a cathodic transfer, leads to varieties of activated oxygen specie, powerful attacking hydrocarbon sites by abstraction of hydrogen, leading to peroxides, and carbon radicals. The carbon radical itself starts to stabilize by abstraction of hydrogen leaving an alkene as new product [10]. (Figure 64)

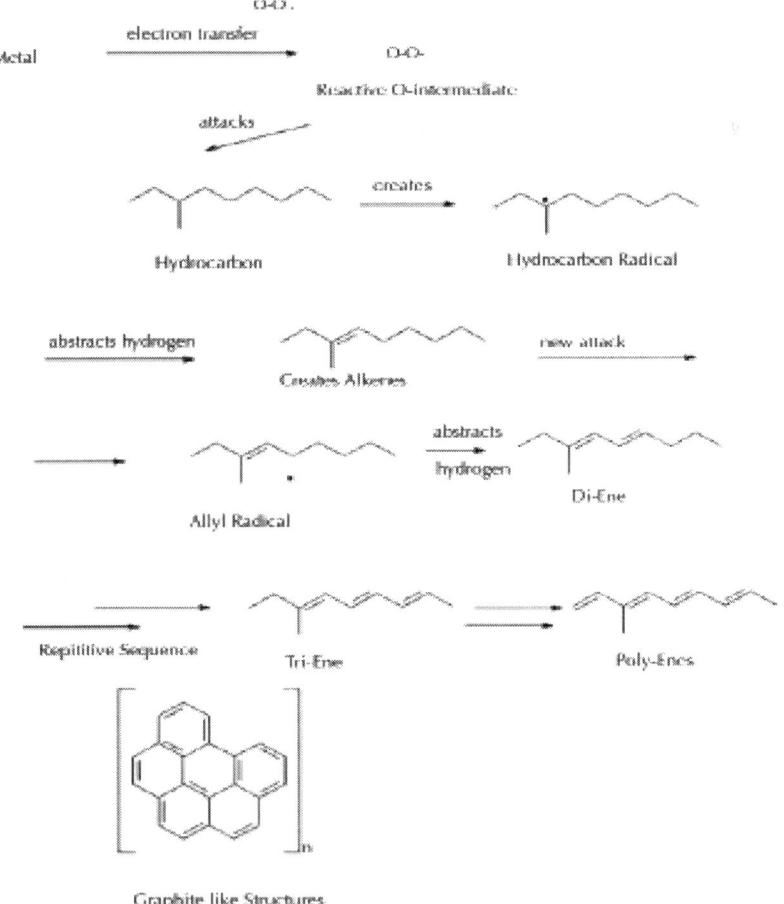

Figure 64: Oxygen – Hydrocarbon Attack sequence.

Due to radical stabilization the new formed alkene starts to continue the oxidation by sequential abstraction of hydrogen, forming di-, tri- and polyalkenes, but also benzene rings. Apart from the hydrogen abstraction, also oxidation takes place by attacking carbon radicals by oxygen. At least the products created by such this procedures are carbonyl compounds, e.g. alcohols, ketones, aldehydes, carboxylic acids and sometimes esters. PAO oxidation at metal surfaces, e.g. iron beyond 120°C results in the formation of lactones (esters that come up by internal reaction between an alcohol group and terminal carboxylic group) (Figure 65).

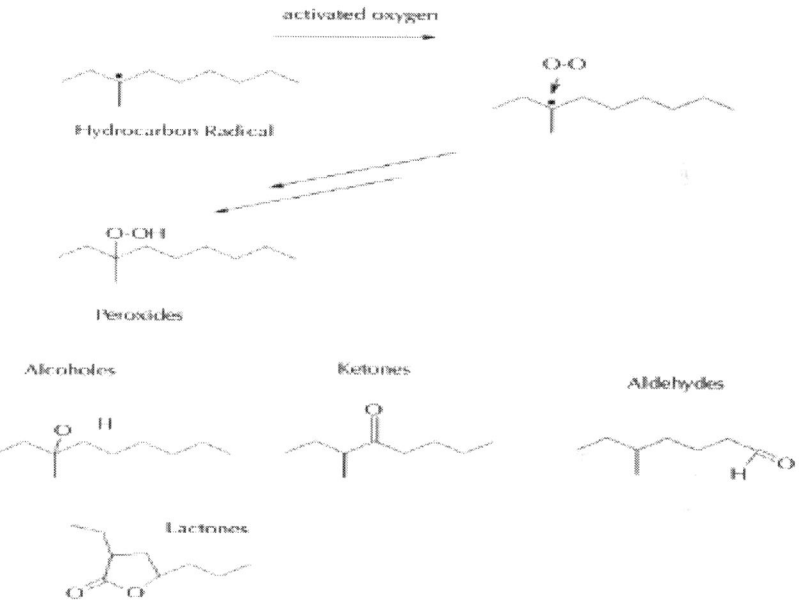

Figure 65: Oxidation sequence of Hydrocarbons toward carbonyl compounds.

Hence oxidation sequences dramatically change the original hydrocarbon chain. If once started it is self accelerating till new, different and stable products are reached. Oxidation is unselective and takes place everywhere in the chain. Hence, plenty of products are formed by radical oxygen assisted processes.

Antioxidants in general prevent the base oil, quenching the oxygen attack by formation of stable radicals. Stabilization of the radicals is realized by a delocalization of the persistent AO radical, created by oxygen attack due to the presence of aromats in the structure (Figure 66).

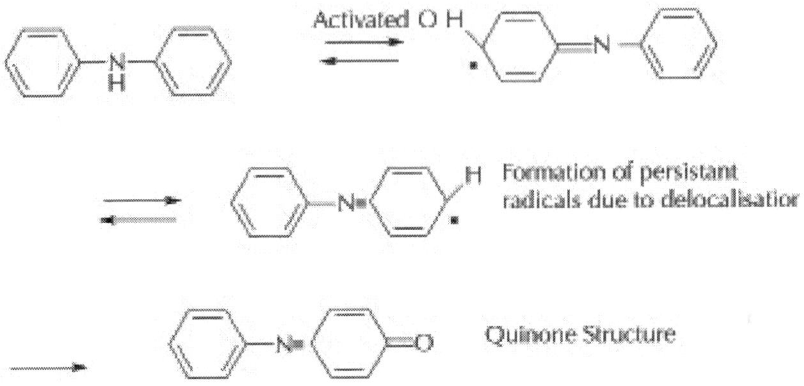

Figure 66: Principal delocalization of radicals created by oxygen attack.

The AO radical subsequently stabilizes to form new products like quinones. The quinone structure may form a dark colored charge transfer complex with the original antioxidant. Very often this causes strong discoloration of AO stabilized lubricants since the charge transfer complexes are very intense in color. Sometimes, for example in the case of polyurea greases, such charge transfer complexes may interfere with the grease structure in terms of solidification.

Persistent radicals formed by AO are dangerous in some cases. In the case of their accumulation in the system they are able to boost oxidation rather than to prevent. Dosage of AO hence should be carefully tested. Formation of either charge transfer complexes or oxidation products by the presence of AO may cause increased formation of sludge in the lubricant if the dosage balance is not appropriate.

Nearly all AO contain aromats as a base principle. Prominent AO candidates are Butylhydroxitoluene (BHT) (A), Alkyldiphenylamine), Phenyl-α-Naphtylamine (PAN) (C) and various others (Figure 67).

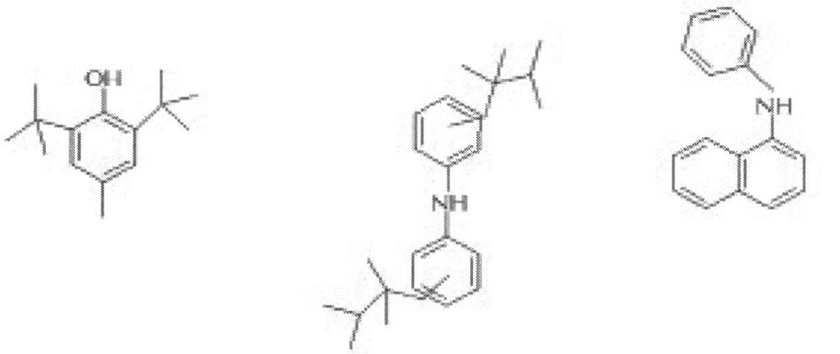

A

Butylhydroxitoluene
BHT

B

Octyldiphenylamine

C

Phenyl-α-Naphtylamine
PAN

Figure 67: Structures of AO: (A): BHT, (B) Alkyldiphenylamine, (C) PAN.

GREASES

General Remarks

Greases are defined apart from their chemical composition by the manufacturing processes. Thickener and oil, getting heated by stirring, start to dissolve. Getting cold, the process of stirring leads to a raw material where amorphous and crystalline structures are merged. The amount of crystals and amorphous materials depends on the nature of the raw materials on the one side and on the rate of heating and cooling on the other side. Rapid cooling causes homogeneous and amorphous structure, as particles are not able to grow to a large size. The raw grease, as effect of the mixture of solid structures has to be homogenized carefully. Homogenization leads to a smoothened appearance of the grease with a scale distribution of thickener particles as effect of the cooling process. Slow cooling generally leads to material with large sized particles as an effect of nucleation and crystal growth. Oil embedding in such structures is different due to the solid structure of the thickener. Stiffness and flowing capability may change as an effect of the merged structure. Greases, even in the case of identical chemical composition may differ significantly by their manufacturing process. Stiffness of greases is defined by the NLGI grade declaration, measured by penetration of standard cone into the grease. The deeper it's penetration the more liquid the grease will be. To get a constant value, the grease is worked by 60 strokes, then tempered to 25 °C and measured by cone penetration. NLGI grades are presented in table 2: [3]

Table 2: NLGI Grades of Greases

NLGI Grade	Cone Penetration in 1/10 mm
000	445-475
00	400-430
0	355-385
1	310-340
2	265-295
3	220-250
4	175-205
5	130-160
6	85-115

Oil Bleeding

Within grease the base oil is bound in different states. Some oil is weakly bound to the thickener nuclei and gets easy released. Oil, bound in micelles and large structures with van-der-Waals and dipolar bonding releases less. Oil release takes place due to centrifugal effects in speeding machinery elements, e.g. bearings, creeping across walls e.g. sealings enhanced by temperature. Successive loss of oil in grease may lead to its change in performance, accompanied by a malfunction. Oil bleeding is measured with different techniques. Within the most popular one the grease is sat on a sieve and pressed by a static load through it a given temperature. Bleeding is measured as a function of time. For bearings the long term bleeding rate should be less than 5 % per weight in 7 days. [3]

Dropping Point

Greases - if heated - start to get liquid at a certain point. Molten grease will leak out at sealing edges and may cause a malfunction of the grease. For bearings the thumb rule is given by dropping point minus 50 °C as the upper point of applicability. [3]

Soap Based Greases

Greases are soft solids, created by a thickener that gelates in suitable base oils. Gelling takes place by intense mixing of thickeners with the base oil, often accompanied by heating till the gelation is reached [3]. (Figure 68):

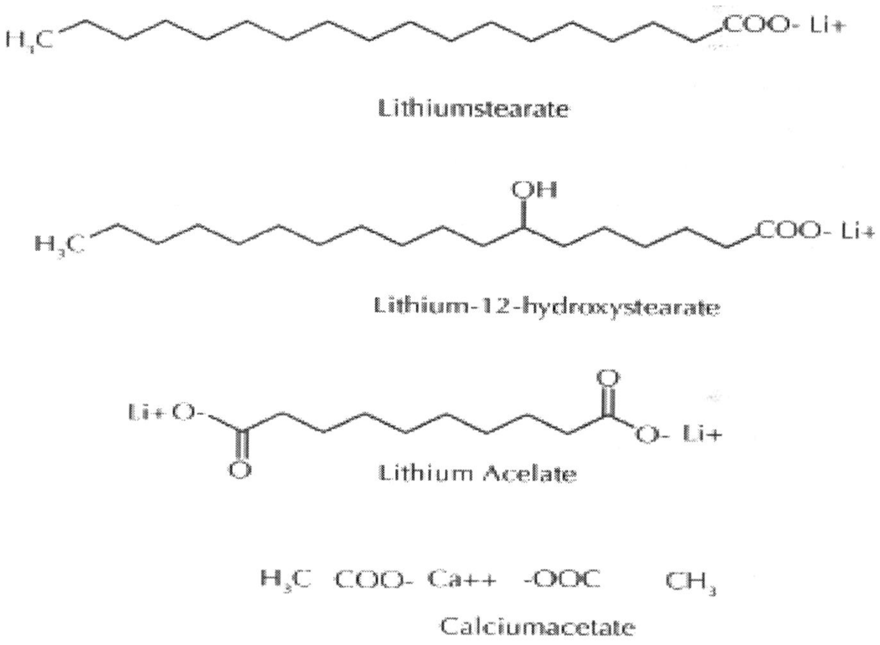

H₃C〜〜〜〜〜〜〜〜〜〜COO- Li+

Lithiumstearate

H₃C〜〜〜〜〜〜〜COO- Li+

Lithium-12-hydroxystearate

Li+ O-〜〜〜〜〜O- Li+

Lithium Acelate

H₃C COO- Ca++ -OOC CH₃

Calciumacetate

Figure 68: Prominent representatives of thickeners for grease production.

Thickeners are all substances where gelling in the base oil is achievable. Prominent representatives are lithium and calcium salts of carboxylic acids, for example Lithium Stearate, Lithium-12-hydroxistearate, Calciumstearate, Calcium-12-hydroxistearate but also Calciumacetate. Lithium Complex Greases are created by the co-existence of lithium-12-hydroxistearate with dicarboxylic acids like Acelaic or sebacaic acid.

Calcium Complex Greases are composed by calcium acetate, Calcium Stearate and calcium-12-hydoxistearate as thickeners.

Salts of magnesium, barium and alumina are used for grease production but to minor extent.

Di and Polyurea Greases (Pu-Greases)

Urea Greases are often called PU-Greases in technical language.

Urea structures are realized by adding amines to isocyanates (Figure 69):

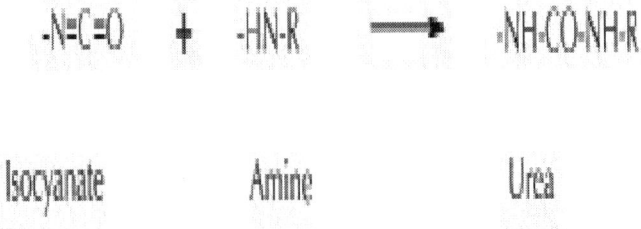

Figure 69: Urea Formation.

Di-Urea grease production take aromatic Isocyanates, like Diphenylmethane Isocyanate (Methylenbisdi-isocyanate, MDI) reacted with various aliphatic amines, like Cyclohexylamine, Alkylamines from C8 to C18 chain length.

Synthesis of the thickeners and grease formation is carried out simultaneously. Ester Oils, like trimellitic acid esters facilitate the synthesis by solving the precursors before the reaction takes place (Figure 70):

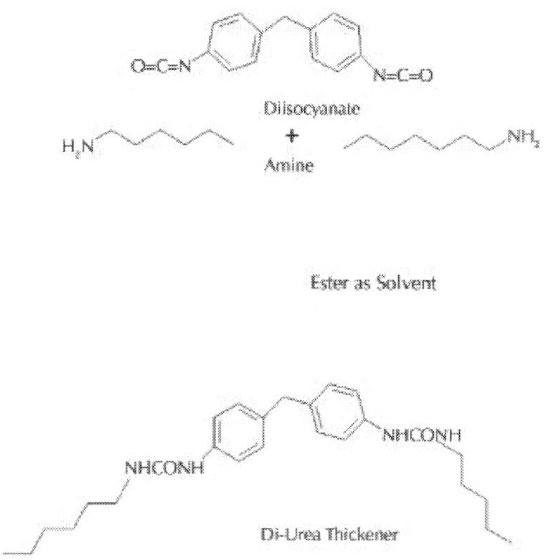

O=C=N⎯⎯⎯⎯⎯⎯ N=C=O

Diisocyanate

+

H₂N⎯⎯⎯⎯⎯⎯⎯⎯ ⎯⎯⎯⎯⎯⎯⎯⎯NH₂

Amine

Ester as Solvent

NHCONH

NHCONH

Di-Urea Thickener

Figure 70: Formation of Di-Urea Grease.

Tetra- and polyurea Greases are created by mixing Di-Isocyanates like MDI or Toluenediisocyanates (TDI) with diamines, like ethylene diamine and monoamines, like Octadecylamine in suitable base oils (Figure 71):

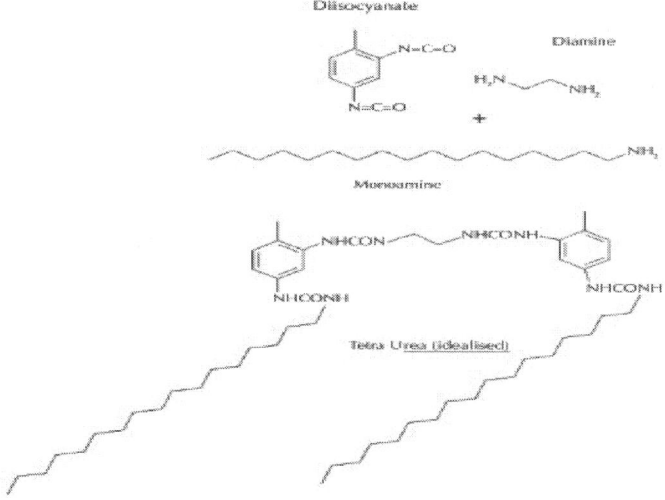

Diisocyanate

N=C=O   Diamine

H₂N⎯⎯⎯NH₂

N=C=O

+

⎯⎯⎯⎯⎯⎯⎯⎯⎯⎯⎯⎯⎯⎯NH₂

Monoamine

NHCON⎯⎯⎯NHCONH

NHCONH   NHCONH

Tetra Urea (idealised)

Figure 71: Formation of Tetra-and Polyurea Greases.

Urea Greases offer plenty nitrogen-hydrogen bridges within their structures. Concordant with the presence of temperature resistant aromatic nuclei and in junction with high temperature resistant base oils, they represent the group of high temperature grease "per se". As to the high variability of taking precursor amines, PU greases offer the possibility to adapt the grease to a given application, much more than soap greases do.

Polyurea Greases that start from tallow amine, tolyenediisocyanate and ethylene di-amine are in accordance with the US FDA regulations H1 (incidental food contact) if H1 base oil (like white mineral oil or PAO) is used. Also the modern EU REACH regulations are valid for polymeric structure.

As the polymeric degree increase, the thickeners may get insoluble and crystalline. Greases are no longer available due to this because a lack of gelling. Due to this fact, variances of PU Greases are restricted.

MDI and especially TDI are ought to be highly toxic by inhalation. Production of PU greases have to take care, than none of the precursors are free in air, nor present in the grease.

Some isocyanates tend to polymerize during production, rather than to react with the amine, especially at the end of the syntheses. Polymeric Isocyanates may remain in the grease and cause severe toxicitiy, especially if the greases are up -heated.

PU Greases are very sensitive toward ingress of OH – groups (e.g. alkalines, water, polyglycoles) as the nitrogen-hydrogen bridging is disturbed. Ingress of such pollutants may cause a change in consistency. Polyglycoles, if heated emit aldehydes that interfere with the NH groups in PU greases. This reaction may end up in making the solid PU liquid! PU greases thus should be monitored to those facts (Figure 72):

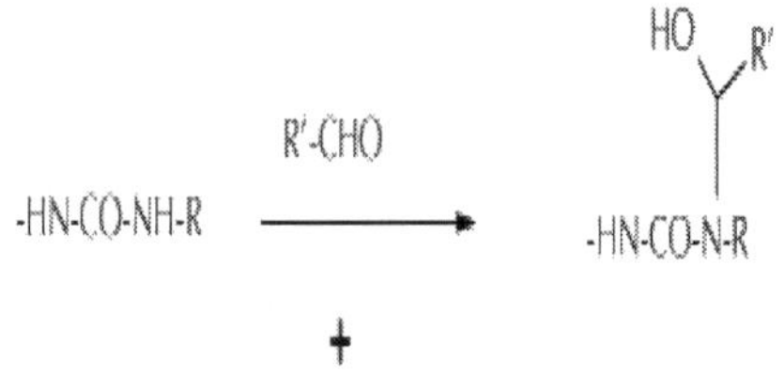

$$ \text{-HN-CO-NH-R} \quad \xrightarrow{\text{R'-CHO}} \quad \underset{\text{+}}{\text{-HN-CO-N-R}} $$

Figure 72: Reaction of PU Grease and Aldehydes.

Other incompatibilities of PU Greases arise from mixtures with clay thickeners due to the presence of either OH (Si-OH) or NH functional groups if the clay is modified by organic amines.

Other Thickeners

Clay Greases-Structure and Use

Clay Thickeners derive from Alumina-Silicates. Due to their high surface and modification they are suitable for gelling base oils, e.g. Esters, Napthenic Base Oils, sometimes Silicones and Phosphoric Acid Esters. Clay Structure is generated by tetrahedral arrangement of Silica with insertion of alumina (see figure) in layers of approximately 1-2 nm distance. Water and other cations may be inserted in the space in between the two layers. Other cations, e.g. magnesium, may also be inserted in between. [3] (Figure 73).

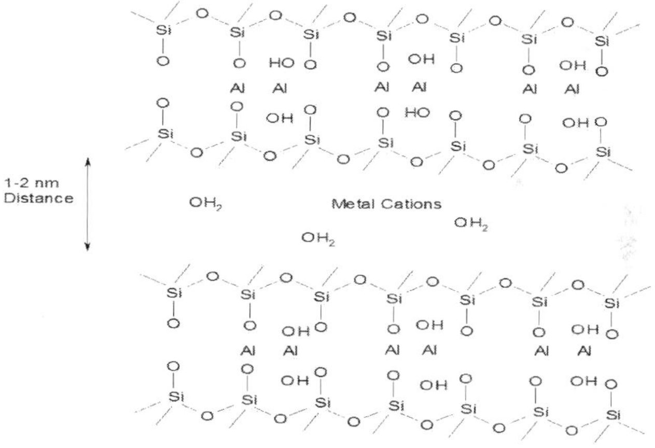

Figure 73: Estimated basic structure of clay.

Gelling takes place by adhesion and insertion of organic molecules in the structure, assisted by polar additives like propylene carbonate. Clay grease is produced by multiple milling the clay with appropriate base oil by addition of water suppliers like glycerol or Propylene carbonate at temperatures below 100°C. If water is lost the structure may break down during the manufacturing. Doing so, the grease produced is a buttery solid with no dropping point.

Use of Clay Greases

Clay greases are used for applications where the grease should not move out and for special high temperature applications, e.g. cement industry in slow motion bearings. Due to the inertness of the inorganic structure toward alkaline and acids, clay greases are preferred in applications where water, alkaline and acids are present, e.g. chain or bearing lubrication with such ingress. Clay is declared as safe for incidental food contact and allowed for lubricants in food industry (USDA H1 regulated) in junction with base oils like white mineral oil, PAO or esters that are allowed for this purpose.

Restrictions in The use of Clay Greases

Restrictions for the use of clay greases are the presence of Lithium, - Calcium or Polyurea Greases that may interfere with the hydrogen bonding of the clay structure. Mixtures of clay and conventional greases should be evaluated very carefully. Clay greases are restricted in bearing lubrication strictly due to over rolling speed. In general the speed factor is limited to ndm (Average of outer and inner diameter of the bearing times the speed (revolution per minute)) of 100.000. Only slow moving bearings could bear clay lubrication.

SILICA

Silica is in use for thickeners as amorphous material, obtained by flame decomposition of Silica Tetrachloride (Figure 74):

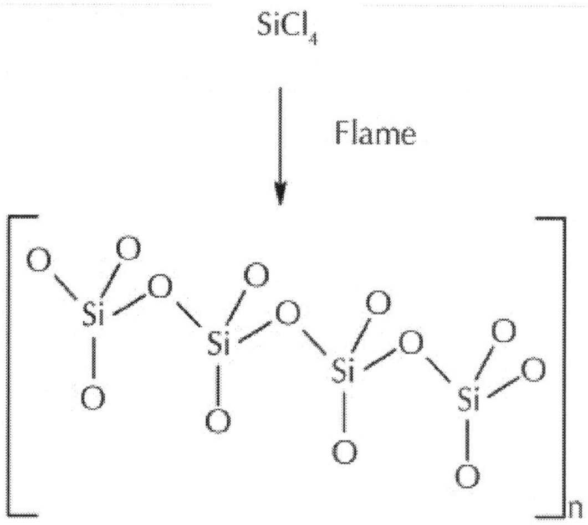

SiCl$_4$

Flame

Figure 74: Principal formation of amorphous SiO2 by flame combustion.

Silica, due to its powerful surface activity may be used as powerful thickener in low percentage for each kind of base oil. Greases obtained by mixing silica with base oils are transparent. The inorganic structure causes no dropping point for such greases. Silica Thickened greases cause steep and irreversible thickening by heating up due to the increase of internal hydrogen bonding. They never should be in use for high speed and high temperature rotating bearings, since they block their motion. The ndm (Average of Bearing Size times revolution per minute) is restricted to 100.000, hence slow motion. Due to the possible entrance of water, silica thickened grease is poorly water stable and should not be in use in applications where water (especially hot water) and alkalines are present. Alkalines react with silica to silicates, starting its degradation.

POLYTETRAFLUOROETHYLENE (PTFE)

PTFE is a convenient thickener in base oils for the purpose of incidental food contact, low friction properties and high temperature. The fluorine entity causes low activity toward oxygen. PTFE Grease is used in oxygen application (valves under oxygen impact), especially with PFPE.

CONCLUSIONS

Tribology is highly guided by physics and chemistry of the lubricants. Functionality of lubricants is given by their physics and their chemical structure. Modern understanding of lubrication hence allows the construction of lubricants appropriate to a given application to a certain extent. Under the conditions of full lubrication their physical properties, e.g. viscosity, viscosity-temperature and viscosity –pressure properties dominate over the chemical structure. Under such circumstances, the lubricant takes away heat (cooling function) from the mating contacts, but also wear and debris (cleaning function). Within a running – in period some reaction layers of lubricant constituents (additives) may be created. Basically those layers stay constant over time and do not change. On the other hand, if lubrication undermines the given roughness's of the mating partners, or overtakes the natural temperature limit given by the restrictions of organic chemistry (e.g. temperatures beyond 150°C), chemistry starts to perform reaction scenario highly related to the nature of the chemical structure of the ingredients in the lubricant. The basic reactions found here are radical reactions, as a fact of the presence of oxygen and iron. Within such radical reaction sequences hydrogen is abstracted, alkenes and alkynes are formed and their oxidation products (aldehydes, ketones, carboxylic acids and their derivatives). Additives, in general improve the lubricants by expanding their limits.

In general, lubrication fundamentals in tribology have overcome the alchemy of the past by numerous efforts taken by the scientific community.

REFERENCES

1. Rudnik L.R., editor. Synthetics, Mineral Oils, and Bio-Based Lubricants. Boca Raton: CRC Press; 2005.

2. Dresel W., Mang T., editors: Lubricants and Lubrication. 2nd Edition. Weinheim: Wiley-VCH; 2007.

3. Klamann D. Schmierstoff und verwandte Produkte. Weinheim: VCH-Verlag; 1982.

4. Mortier R.M., Fox M.F., Orszullik T.M., editors. Chemistry and Technology of Lubricants Dordrecht: Springer; 2010. http://link.springer.com/book/10.1007/978-1-4020-8662-5/page/1 (accessed 27 December 2012).

5. Dowson D., Taylor C., Childs T., Dalmaz G. editors. Lubricants and Lubrication. In: Tribology Series 30 : Proceedings of the 21st Leeds-Lyon Symposium on Tribology. Amsterdam : Elsevier; 1995.

6. Bloch, H.P., Practical Lubrication for Industrial Facilities. Lilburn: Fairmont Press; 2000.

7. Stepina V., Vesely V. Lubricants and Special Fluids. Amsterdam: Elsevier; 1992.

8. Lansdown A.R., Lubrication and lubricant selection: a practical guide. 3rd Edition. John Wiley & Sons; 2004.

9. Rudnick L. R., editor. Lubricant Additives: Chemistry and Applications, 1st Edition. New York:Marcel Dekker, 2003.

10. March, J., Advanced Organic Chemistry: Reactions, mechanisms, Structure. New York: Wiley-VCH; 1992.

Chapter 7

Quantitative Evaluation of Completion Techniques on Influencing Shale Fracture 'Complexity'

N. Nagel[1], F. Zhang[1], M. Sanchez-Nagel[1], and B. Lee[1]

[1] Itasca Houston, Inc., USA

ABSTRACT

In many of the active shale plays, the extremely low permeability of the shale means simple, bi-planar hydraulic fractures do not provide enough surface area to make an economic well. In these cases, the optimal, economic completion requires stimulation of the natural fracture system - often called increasing the 'complexity' of the stimulation.

A number of different multi-well completion techniques have been proposed to enhance shale complexity. The 'simul-frac' technique is where companion wells are stimulated at the same location at the same time, whereas the 'zipper-frac' technique employs companion wells that are stimulated in staggered locations at the same time. The intention with these techniques is to alter either or both the stress field and the pore pressure field to enhance the shearing of natural fractures.

In this paper, we present the results of a numerical study to quantitatively evaluate the effectiveness of multi-well completion techniques, particularly the 'modified zipper-frac' technique, to optimize shale completions. The study includes a parametric study of the effects of in-situ stress conditions, natural fracture orientation and fracture friction, and hydraulic fracture layout on changing near and far-field natural fracture shear (complexity). Changes in the stress field, particularly shear stress, are considered the primary means of increasing fracture complexity. The quantitative results of the study provide a means to optimize the application and design of different multi-well completion techniques as a function of the presented parameters. Optimized completion designs mean lower well costs, greater production and, ultimately, improved well economics.

INTRODUCTION

Much has now been written about the boom in shale gas and shale oil developments in the United States and around the world. In its recent assessment for example, the Energy Information Agency (EIA 2012) noted that North Dakota has become the second largest oil producer in the United States due to production from the Bakken shale. In addition, the EIA (EIA 2013) has predicted that the United States will continue to add more than 230,000 bpd of oil production per year through the end of the decade and become a net exporter of natural gas within the decade. Expenditures on shale gas and shale oil developments have also rapidly increased. For example, more than $54 billion dollars was spend in drilling and development operations in the seven major US shale developments in 2012 (Clover Global Solutions 2012), with the bulk being spent in the Eagle Ford and Bakken plays.

Shale developments, notably beginning in the Barnett in the 1990s, have been driven by: 1) the application of horizontal wells; 2) the application and improvements in hydraulic fracturing; and 3) significant commodity prices (GWPC 2009 and King 2010). Because of the low permeability in most shale developments (nano-darcy permeability in shale gas plays and micro-darcies in shale oil plays), hydraulic fracturing is a key technology because, as noted by King (2010), the presence of, and the ability to open and maintain flow in, both the primary and secondary natural fracture systems is critical. King further noted the importance of maximizing the fracture-to-shale contact area and optimizing the development, placement, and length of small fractures to enhance and stabilize well production (i.e., optimizing the stimulation of the natural fracture system - that is, increasing natural fracture 'complexity').

Because the stimulation of the natural fracture system is critical to many shale developments, a number of different multi-well completion schemes have been devised in an effort to improve the ability to enhance the stimulation of natural fractures. Three of the common completions schemes are shown inFigure 1. In simultaneous fracturing (plot A in Figure 1), the concept is that hydraulic fracturing both wells at the same time enhances the stimulation of the natural fractures. In the sequential (zipper) frac concept (plot B), the residual stress field from well #1 is thought to enhance the stimulation of the natural fractures when well #2 is stimulated. Finally, in the modified zipper-frac concept (plot C,Figure 1), the sequential stimulation of offsetting stages is thought to enhance the stimulation of the natural fractures.

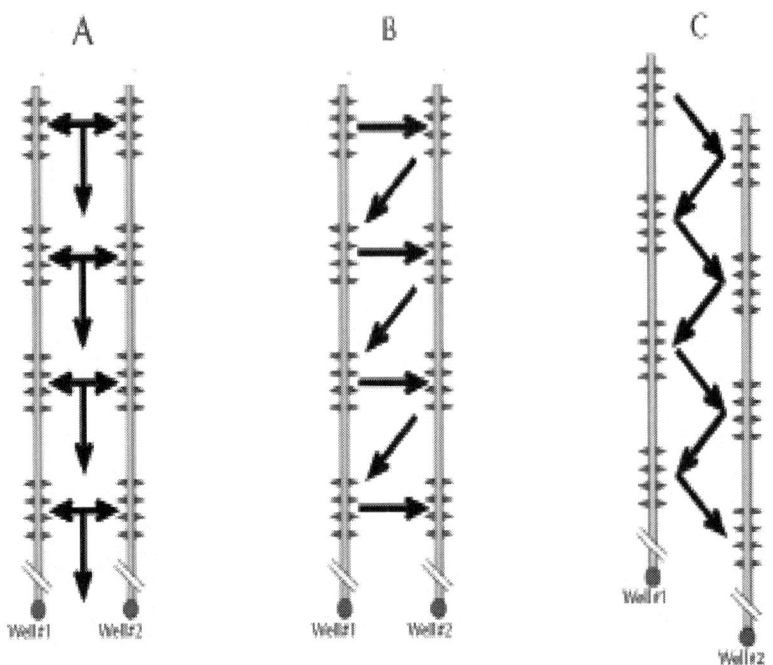

Figure 1: Common shale completion schemes. A) Simultaneous hydraulic fracturing; B) Sequential fracturing (zipper-frac); and C) Modified zipper-frac.

Natural Fracture Behavior

A critical component to understanding the efficacy of multi-well completion techniques on increasing shale complexity is the understanding of the geomechanical behavior of natural fractures. The authors have written extensively about the mechanical behavior of natural fractures and the results of numerical modeling (both continuum and distinct element modeling) of the response of natural fractures to hydraulic fracture stimulation (Nagel et al. 2012a, Nagel et al. 2012b, Nagel et al. 2012c, Nagel et al. 2011a, Nagel et al. 2011b, and Nagel and Sanchez-Nagel 2011). Of first interest in evaluating the impact of multi-well completion schemes on the stimulation of natural fractures is the basic behavior of natural fracture shear and deformation.

Nagel et al. (2012c) summarized five 'conditions' for natural fracture shear to occur:

- The shear stress along the fracture grows to exceed the shear strength with no change in fracture friction, fracture normal stress, or fracture pore pressure;
- Due to thermal or chemical changes, fracture friction is reduced while the shear stress along the fracture is unchanged and the fracture normal stress and fracture pore pressure are unchanged;
- The fracture normal stress decreases with no change in the shear stress along the fracture, the fracture friction coefficient, or fracture pore pressure;
- The fracture pore pressure increases with no change in the shear stress along the fracture, the fracture friction coefficient, or fracture normal stress; and
- A variety of combinations of the above.

Of these, conditions 3 and 4 (and, by default, condition 5) are believed to be most relevant to the behavior of fractured shale plays during hydraulic fracturing. The impact of these conditions is shown graphically in Figure 2. Figure 2 is a schematic representation of the results of a direct shear test on a fractured rock sample. The x-axis represents the shear displacement along the fracture during the test, and the y-axis represents the shear stress imparted to the rock in order to achieve the given displacement. Four stress-displacement profiles are shown, which represent increasing effective normal stress on the fracture. As the effective normal stress is increased, both the peak shear stress necessary to initiate non-elastic displacements and the shear stress necessary to continue non-elastic displacements increase.

The implications of this behavior are critical to understanding the behavior of natural fractures during hydraulic fracturing. As shown in Figure 2, as the normal stress acting on natural fractures increases (due, for example, to the inflation of an induced hydraulic fracture), greater shear stress is required to cause shear slippage and displacement along a natural fracture. Effectively, increasing the normal stress stabilizes the natural fractures. At the same time, as pressure increases within a natural fracture (due, for example, to bulk fluid flow into the natural fractures or pressure diffusion from the induced hydraulic fracture), less shear stress is required to cause shear slippage. Given this behavior for natural fractures, and the goal of increasing the shear stimulation of these during hydraulic fracturing, the evaluation of the impact of completion scheme on well stimulation should focus on whether or not the completion scheme increases the shear of the natural fractures.

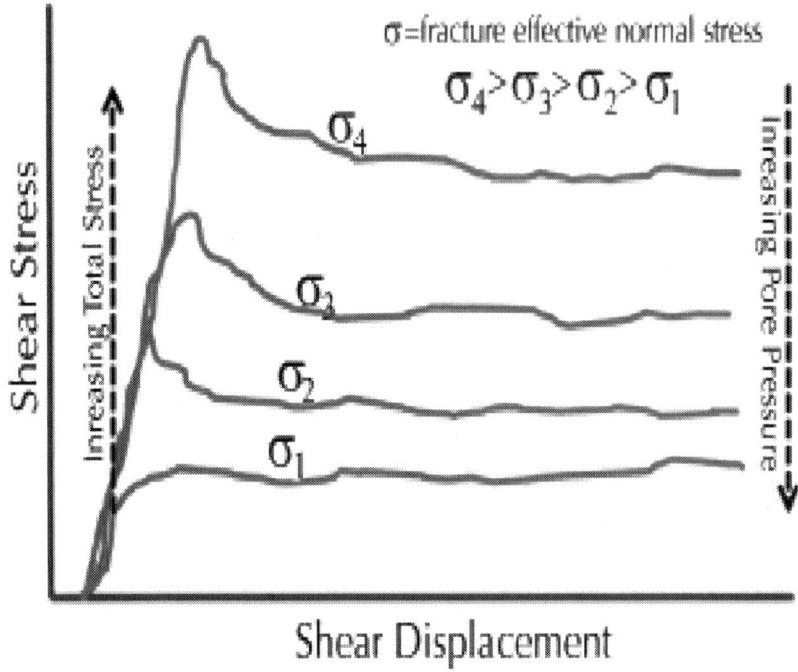

Figure 2: Shear-displacement profiles as a function of normal effective stress from direct shear testing of fractured rock.

Hydraulic Fracturing and Stress Shadows

If increasing normal stress stabilizes natural fractures, then evaluating the stress changes from a hydraulic fracture is a required element of evaluations to optimize shale complexity. As far back as Sneddon's work on the evaluation of stress near a crack (Sneddon 1946), numerous authors have looked at the impact of stress field changes around hydraulic fractures (Nagel and Sanchez-Nagel 2011 and Warpinski et al. 2012). The stress field change, principally the increase in the minimum horizontal stress, Shmin, caused by a hydraulic fracture (typically the final, propped hydraulic fracture) is called the stress shadow effect or simply the stress shadow. Figure 3 shows the stress shadow (increase in Shmin) from a single hydraulic fracture that was 300m long and 140m in height (along the x-z plane on the right side of the model) in a model that is 1000m cube.

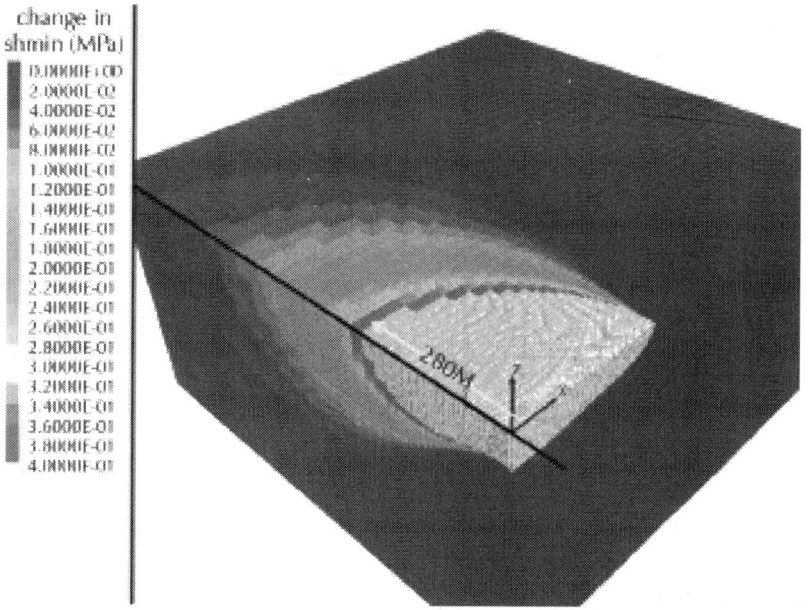

Figure 3: Stress shadow contours from a single 300m long/140m high hydraulic fracture with a 5MPa net pressure applied on the x-z plane. The cutaway image was created by cutting along the y-z and x-y planes. The model shown is 1000m in each of the x and y-directions. The white area is a region of stress change greater than the color scale shown (from Nagel et al. 2013).

As shown, note both the long distance over which the stress change occurs – to the edge of the 1000m long block simulated – and the vertical spreading with distance. At large distances, the change in stress is seen to affect a total formation height more than double the original height of the created fracture. Note also the near-complete lack of stress change beyond the horizontal tip of the hydraulic fracture. Overall, the following can be summarized about stress shadows (Nagel and Sanchez-Nagel 2011):

- The increase in Shmin (stress shadow) extends significant distances behind a fracture and spreads out above and below the fracture but not beyond the tip of the fracture.

- The increase in Shmin due to a hydraulic fracture is largely unaffected by either the in-situ rock mechanical properties or the stress ratio (though these do appear to affect changes in the vertical stress and the SHmax stress).

- A horizontal shear stress field occurs with the fracture tip and does not extend back to the wellbore. This suggests that, at some distance behind the fracture tip, the effect of the stress shadow is to stabilize the natural fracture system.
- Reducing fracture spacing results in a greater minimum Shmin stress increase in the inter-fracture region as the stress shadows from each fracture overlap more with reduced fracture spacing.

Natural Fracture Behavior and Stress Shadows: Implications for Completion Strategies

The combined consideration of the basic mechanical behavior of natural fractures and the nature of stress shadows suggests the following for a multi-well completion strategy:

- The stress shadow effect, that is the increase in the principal stresses around a hydraulic fracture, causes a stabilization of natural fractures. This can only be overcome by increasing the fluid pressure within the natural fractures (suggesting a desire to increase the net pressure, which would also increase the stress shadow). Decreasing stage spacing, or overlapping hydraulic fractures from different wells, will tend to increase the stress shadow effect and impair the stimulation of natural fractures.
- Because the stress shadow effect does not extend horizontally beyond the tip of the hydraulic fracture (the x-direction in Figure 3), when two fractures are simultaneous created from parallel wellbores, the fractures will not 'see' each other until the tip regions are very near to each other (and increase the potential for screenout during a stimulation).

Numerical Simulation of Completion Strategies: Modified Zipper-Frac

In this paper, numerical simulation results are presented for the evaluation of the modified zipper-frac multi-well completion strategy. The simulations were conducted with a 2D discrete element model (DEM) under different well configurations for two different natural fracture networks, different fracture friction angles, and different stress ratio conditions.

MODEL SETUP AND SIMULATION MATRIX

2D Dem Model Capabilities

A two-dimensional DEM code was used in all the simulations presented. The code used was a general-purpose program based on the distinct element method for discontinuum modeling. The code can simulate the response of discontinuous media (such as a jointed rock mass) subjected to either static or dynamic loading. The discontinuous medium is represented as an assemblage of discrete blocks, and discontinuities are treated as boundary conditions between blocks. Large displacements along discontinuities and rotations of blocks are allowed. Individual blocks behave as either rigid or deformable material. Deformable blocks are subdivided into a mesh of finite-difference elements, and each element responds according to a prescribed linear or nonlinear stress-strain law. The relative motion of the discontinuities is also governed by linear or nonlinear force-displacement relations for movement in both the normal and shear directions. The basic formulation of the code assumes a two-dimensional plane-strain state. This condition is associated with long structures or excavations with a constant cross-section acted on by loads in the plane of the cross section. Discontinuities, therefore, are considered as planar features oriented normal to the plane of analysis. For plane-strain analyses, blocks may exhibit plastic yield, and failure can occur in the out-of-plane direction if the out-of-plane stress becomes a major or minor principal stress.

The critical modeling features for the simulation of hydraulic fracturing include:

- A rock mass is modeled as an assemblage of rigid or deformable blocks. The size, shape, and orientation of the blocks are defined by the imported Discrete Fracture Network (DFN) or by the internal fracture generator.

- Discontinuities are regarded as distinct boundary interactions between blocks, and continuous and discontinuous joint patterns or joint properties can be generated on a statistical basis or from an imported DFN.

- Fractures are created within, and propagate along, the static block boundary planes; however, propagation can be modeled explicitly based upon the stress intensity factor at the fracture tip. Fracture behavior is prescribed by the block interactions. Thus, natural fracture aperture is, for example, affected by shear displacement and fracture fluid pressure.

- Fluid flow is limited to flow within the fractures, and matrix fluid (and, for example, fluid leakoff) is not considered.

Model Setup

Figure 4 shows the setup and dimensions of the 2D model in planview at the centerline of the horizontal wellbores (located along the left and right sides of the model shown). Table 1 summarizes the model mechanical parameters while Table 2 summarizes the stress conditions used. The total model was 1200m long in the direction of Shmin (vertical or y-direction) and 225m wide in the direction of SHmax (horizontal or x-direction) as shown in plot A of Figure 4. In order to avoid boundary effects, the vertical boundaries were placed at a large distance (> 550m) from the simulated hydraulic fractures and roller boundaries were applied. The horizontal boundaries were considered to be symmetry planes at the wellbore locations (as only half the fracture length was modelled) and roller boundaries were also applied.

Two different natural fracture patterns were employed. In plot B of Figure 4 (note that plot B and C represent the central core in green from plot A), the '180°' fracture pattern is shown. This pattern contains two fracture sets, which are nominally orthogonal to each other and aligned with the principal stress directions. The second fracture pattern, called the '145°' pattern, is shown in plot C. For the 145° pattern, the same two fracture sets from the 180° pattern have been rotated roughly 45° relative to the principal stresses.

The simulated hydraulic fractures are shown in solid and dashed black lines in plots B and C. The solid line represents the first hydraulic fracture location (Xf1) and the dashed lines represent the location of the second hydraulic fracture (Xf2) at a distance of 20m, 35m, and 45m offset along the wellbore from Xf1. When fully propagated, Xf1 and Xf2 were 125m long (their fracture half length).

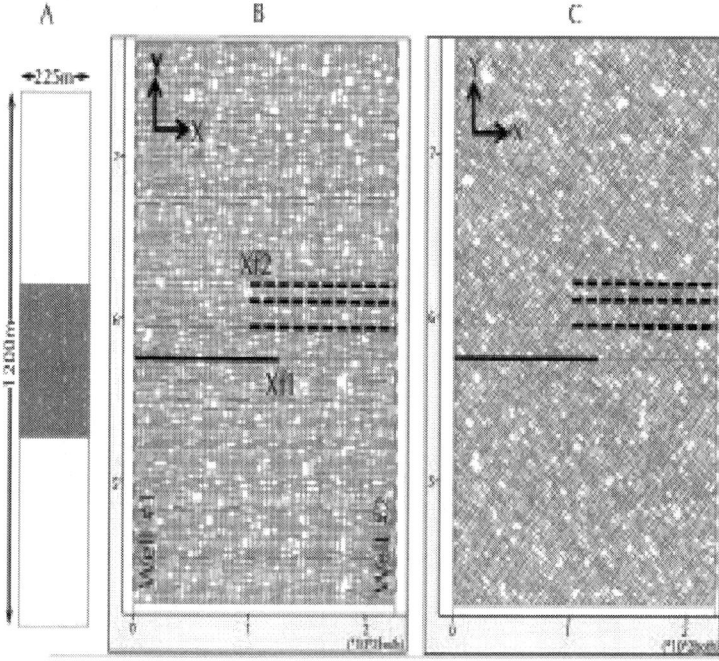

Figure 4: DEM model setup and dimensions. A) Full model - only the middle section in green contained fractures; B) Middle section natural fracture pattern for the '180°' model; and C) Middle section natural fracture pattern for the '145°' model. The location of simulated hydraulic fractures are represented by the black lines. Horizontal wellbores are located along the full length of the left and right sides of the model.

Table 1: Mechanical Parameters Used For Model Construction

	DFN #1	DFN #2
Case Name	'180°'	'145°'
Matrix Young's modulus	27.6 GPa	27.6 GPa
Matrix Poisson's ratio	0.25	0.25
Fracture Set #1 Orientation	N180°	N145°

Set#1 Trace Length, mean	35m	35m
Set#1 Trace Length, st. deviation	10m	10m
Set#1 Gap Length	5m	5m
Set#1 Spacing, mean	2m	2m
Set#1 Spacing, st. deviation	0.75m	0.75m
Fracture Set #2 Orientation	N90°	N45°
Set#2 Trace Length, mean	35m	35m
Set#2 Trace Length, st. deviation	10m	10m
Set#2 Gap Length	5m	5m
Set#2 Spacing, mean	3m	3m
Set#2 Spacing, st. deviation	1m	1m
Fracture Normal Stiffness	2e11 Pa	2e11 Pa
Fracture Shear Stiffness	2e11 Pa	2e11 Pa
Initial Fracture Aperture	0.1 mm	0.1 mm

Table 2: Model Stress And Pore Pressure Data

	DFN #1		DFN #2
Case Name	'180°'		'145°'
Vertical Stress, Sv	55.2 MPa	55.2 MPa	55.2 MPa
Max. Horizontal Stress, SHmax	44.8 MPa	44.8 MPa	44.8 MPa
Min. Horizontal Stress, Shmin	37.9 MPa	43.5 MPa	37.9 MPa
Pore Pressure	27.6 MPa	27.6 MPa	27.6 MPa

Modeling Assumptions

For any numerical modeling, assumptions need to be made for the problem being simulated. For the simulations described in this paper, the following assumptions were made:

- 2D, plane strain condition exists (i.e., effects above and below the vertical centerline of a vertical hydraulic fracture do not impact the results).

- Hydraulic fracturing is a quasi-static process and both fracture propagation and injection rate effects can be ignored for the cases being simulated (i.e., the hydraulic fracturing process can be represented by static simulations of the fracture at specific lengths under a given net pressure).

- Events within the formation at the tip of the hydraulic fracture can be simulated without fluid flow within the natural fractures (i.e., the behavior within the formation at the tip is dominated by the changes in total stress and pore pressure changes are negligible).

- Net injection pressure was a constant 4 MPa within the hydraulic fractures Xf1 and Xf2 for all simulations.

- Simulations were conducted first for single hydraulic fractures (without influence from a second, nearby hydraulic fracture). Then simulations were conducted for dual hydraulic fractures and compared to results from two hydraulic fractures acting independent of each other.

Simulation Matrix

In total, nearly 100 simulations were performed in order to explore the behavior of the modified zipper-frac completion scheme. 20 simulations were performed to look at the shear results from a single hydraulic fracture with the '180°' (DFN#1 from Tables 1 and 2) varying the length of both fractures Xf1 and Xf2 separately (Xf1 represents the left-side hydraulic fracture - the solid black line in Figure 4 - and Xf2 represents the right-side hydraulic fracture – the dashed lines in Figure 4) from 25m to 125m in 25m increments with a friction angle of 15 degrees (10 simulations) and repeating these with a 25 degree friction angle (10 simulations). Then 60 simulations were performed to look at the efficacy of the modified zipper-frac completion by performing simulations with dual hydraulic fractures with both DFN models (the '180°' and '145°') varying the length of Xf2 (0m, 50, 75m, 100m, and 125m) for a constant Xf1 of 125m length for three separation spacings (20m, 35, and 45m, where separation is the horizontal offset of the Xf1 and Xf2 fractures as shown in Figure 4) and for two friction angles (15 degrees and 25 degrees in the '180°' model and 25 degrees and 35 degrees in the '145°' model). Finally, an additional 15 simulations were performed with the '180°' model varying the initial in-situ stress (see Table 2) and Xf2 length, and keeping the friction angle at 25 degrees.

QUANTITATIVE NUMERICAL EVALUATION OF MODIFIED ZIPPER-FRACS

Natural Fracture Shear from a Single Hydraulic Fracture

A first series of base case simulations were conducted in order to evaluate the natural fracture shear from a single fracture. The simulations looked at the growth of the Xf1 hydraulic fracture as well as the Xf2 hydraulic fracture for two different fracture friction angles. These base case simulations are important because, in order to correctly evaluate the benefit or detriment of the dual frac modified zipper-frac completion, the effect of the two fractures Xf1 and Xf2 completely independent of each other needs to be considered.

Figure 5 shows the natural fracture shear region (in blue) for both the 15° (plot A) and the 25° friction angle simulations for the '180°' DFN model when Xf1 had a fracture half-length of 100m. As an example, the total cumulative length of natural fractures at shear condition in plot A (the sum of the length of the natural fractures in blue in Figure 5) was 300.1m versus 80.8m in plot B. Figure 6 shows the combined shear for the five Xf1 length simulations and the combined sheared area (shaded area) for the propagation of Xf1 from the wellbore to a 125m half-length. As expected, the area of shear for the lower friction simulations is considerably greater (5740m²) than for the higher friction simulations (2220m²).

The shaded area in Figure 6 and others, adjusted for the length of Xf2 in the dual fracture simulations, represents the sheared area for Xf2 when Xf2 was created independently of Xf1. This shaded area then allows for comparison of independent Xf1 and Xf2 hydraulic fracture effects to modified zipper-frac effects.

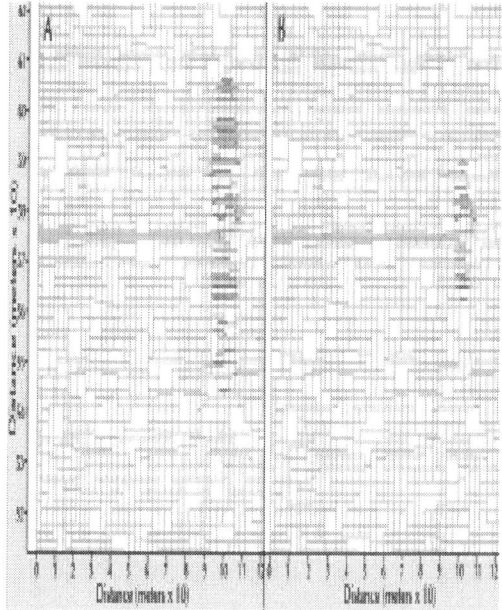

Figure 5: Natural fracture shear (blue lines) for a 100m-long Xf1 hydraulic fracture. A) Shear for the 15° fracture friction case; and B) Shear for the 25° fracture friction case.

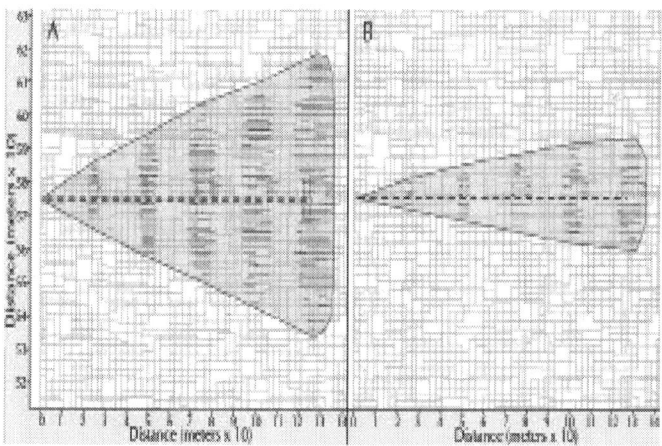

Figure 6: Cumulative natural fracture shear (shaded area) from simulations at 25 to 125m hydraulic fracture half length. A) Shear for the 15° friction case

with an area of 5740m²; and B) Shear for the 25° friction case with an area of 2220m².

Figure 7 shows the growth of sheared natural fracture length as a function of hydraulic fracture half-length for the 15° and 25° natural fracture friction cases for all 20 single fracture simulations. Not surprisingly, given the slight variability in the statistics for natural fracture generation, there are slight, insignificant differences between the results for the Xf1 and Xf2 simulations.

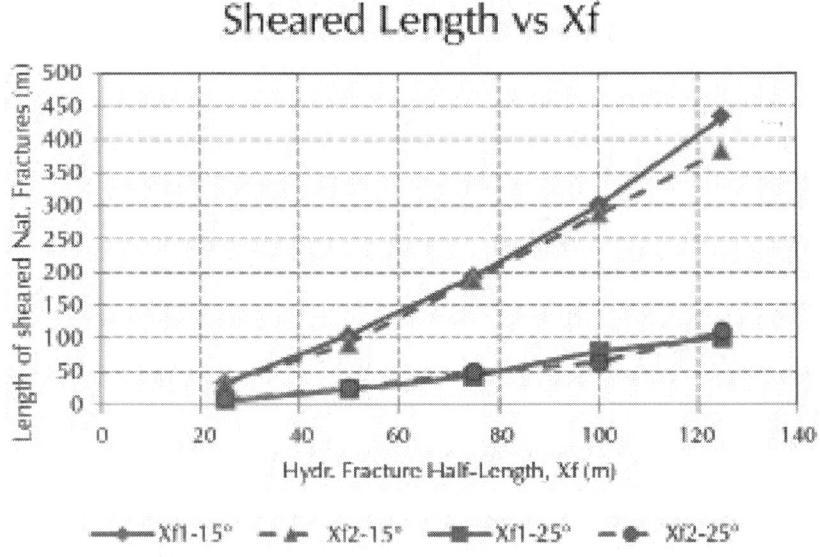

Figure 7: Cumulative natural fracture shear length versus hydraulic fracture half-length for single hydraulic fracture simulations.

A similar evaluation to Figure 6 was performed for the '145°' DFN case as shown in Figure 8. Note that in plot A, natural fracture friction was 25° and in plot B natural fracture friction was 35°.

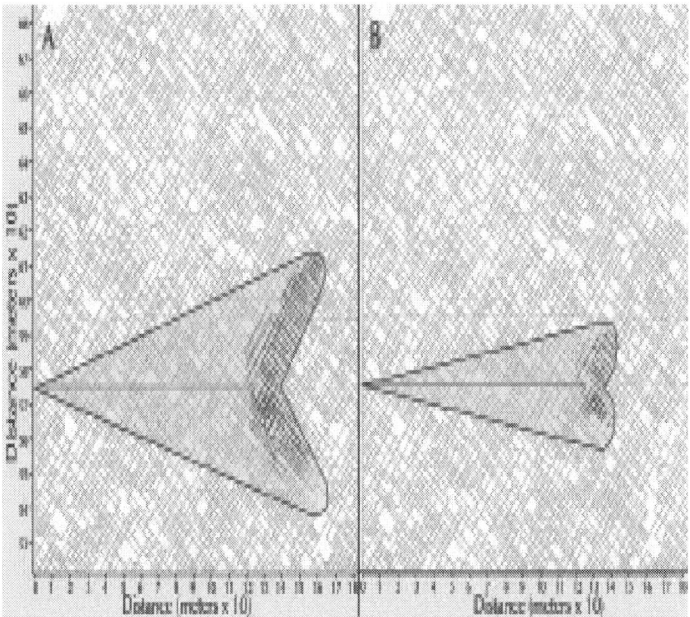

Figure 8: Cumulative natural fracture shear (shaded area) from simulations with the '145°' DFN. A) Shear for the 25° friction case with an area of 5250m²; and B) Shear for the 35° friction case with an area of 2490m².

In summary, Figures 5 through 8 suggest the following:

- Natural fracture shear from the total stress change caused by the inflated hydraulic fracture travels with the tip of a growing hydraulic fracture (as reported in Nagel et al. 2011a and 2012a).
- The length of natural fractures being sheared increases significantly with length (Figure 7).
- The length of natural fractures being sheared is strongly a function of natural fracture friction angle.
- The area (and by default the volume) of formation sheared by a single fracture can also be significant (5740m² for the 15° case and 2220m² for the 25° case of the '180°' DFN and 5250m²for the 25° case and 2490 m² for the 35° case of the '145°' DFN).
- The orientation of the natural fractures significantly affects natural fracture shear for a given fracture friction (at 25° friction, more than twice the shear occurred for the '145°' DFN as for the '180°' DFN).

Natural Fracture Shear Superimposing Two Hydraulic Fractures

Figures 9 through 12 show the superimposed natural fracture shear areas from independent hydraulic fractures for multi-well completions with hydraulic fracture separations ranging from zero (equivalent to either the simultaneous or zipper-fracs) to 45m for both fracture friction cases.

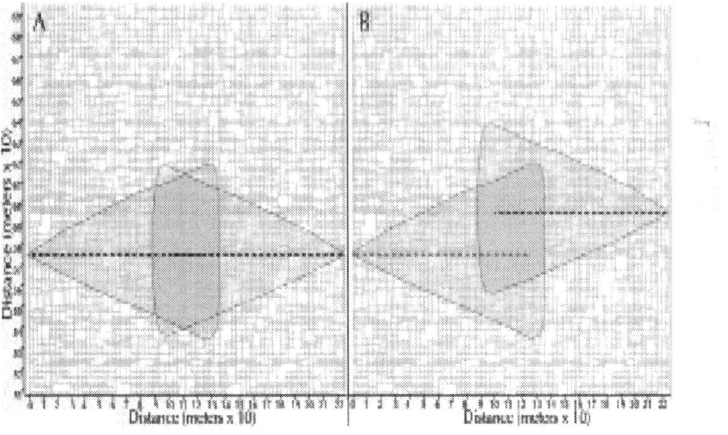

Figure 9: Superimposed natural fracture shear areas for the 15° friction case when Xf1 and Xf2 are both 125m in half-length. A) Zero separation between the two hydraulic fractures; and B) A 20m separation between fractures.

Within the figures, the regions of overlap would likely represent areas of 'wasted' hydraulic fracture shear (and, perhaps, a negative effect on production as excess shear will cause the natural fractures to reclose and even fill with gouge). Ideally, the best effect, assuming no geomechanical interaction between the two hydraulic fractures, may be when the natural fracture shear regions just touch each other (not unlike the situation in Figure 12A).

Figures 9 and 10 suggest that overlapping the lengths of the hydraulic fracture (as proposed in the modified zipper-frac completion) creates large overlapping natural fracture shear areas for the 15° fracture friction case. Further, increasing the hydraulic fracture separation out to 45m still results in considerable overlap of the shear regions. In contrast, with the reduction in shear area due to the increase in natural fracture friction in the 25° friction case in Figures 11 and 12, the shear region overlap goes away at a 35m hydraulic fracture spacing, and for the 45m separation case an unsheared region (Figure 12, plot B) occurs between the hydraulic fractures.

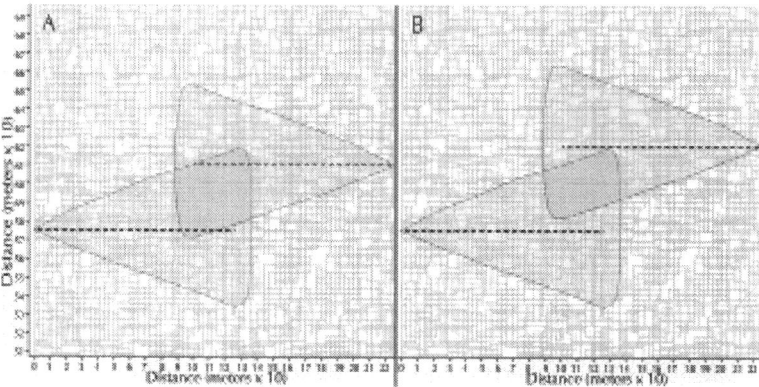

Figure 10: Superimposed natural fracture shear areas for the 15° friction case when Xf1 and Xf2 are both 125m in half-length. A) A 35m separation between the two hydraulic fractures; and B) A 45m separation between fractures.

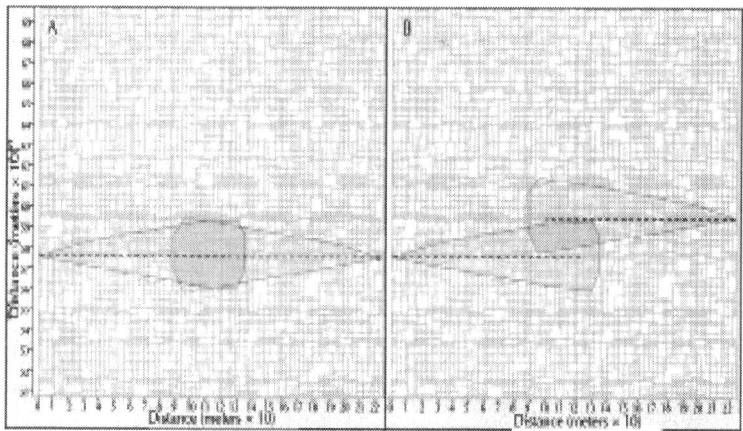

Figure 11: Superimposed natural fracture shear areas for the 25° friction case when Xf1 and Xf2 are both 125m in half-length. A) Zero separation between the two hydraulic fractures; and B) A 20m separation between fractures.

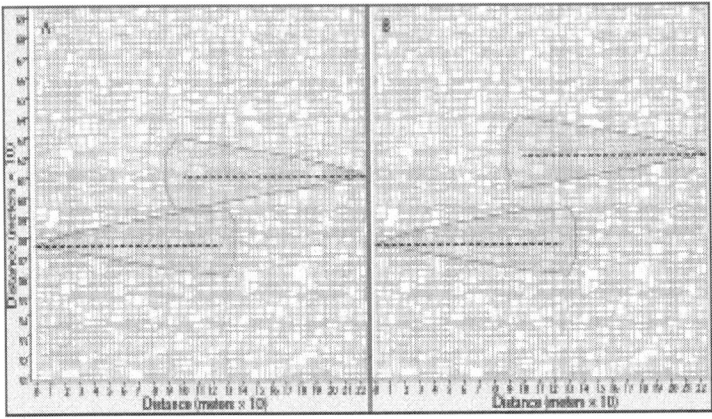

Figure 12: Superimposed natural fracture shear areas for the 25° friction case when Xf1 and Xf2 are both 125m in half-length. A) A 35m separation between the two hydraulic fractures; and B) A 45m separation between fractures.

Natural Fracture Shear from Dual, Competing Hydraulic Fractures

Shear Results for the '145°' DFN and 20m Hydraulic Fracture Separation

Figures 13 through 20 show the generation of natural fracture shear from combinations of the two hydraulic fractures Xf1 and Xf2 as a function of Xf2 length and natural fracture friction for the '145°' DFN with a hydraulic fracture separation of 20m. Plot A shows the sheared natural fractures in blue and open fractures in red; plot B shows the same data with an overlay of sheared natural fracture area (similar to Figure 8) as if hydraulic fractures Xf1 and Xf2 propagated independent of each other.

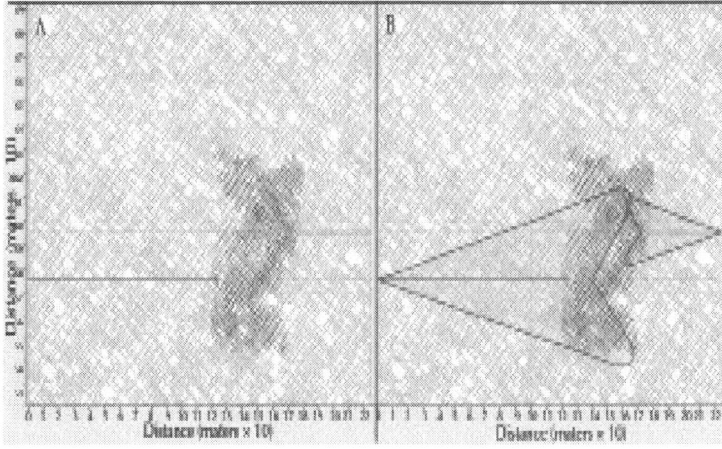

Figure 13: Natural fracture shear in blue from propagating hydraulic fractures Xf1 (from the left at 125m) and Xf2 (from the right at 50m) for a natural fracture friction of 25° and 20m hydraulic fracture separation. Red represents open fractures. A) Shear and open fractures only; and B) Shear and open fractures with overlay of shear area as if Xf1 and Xf2 propagated independently.

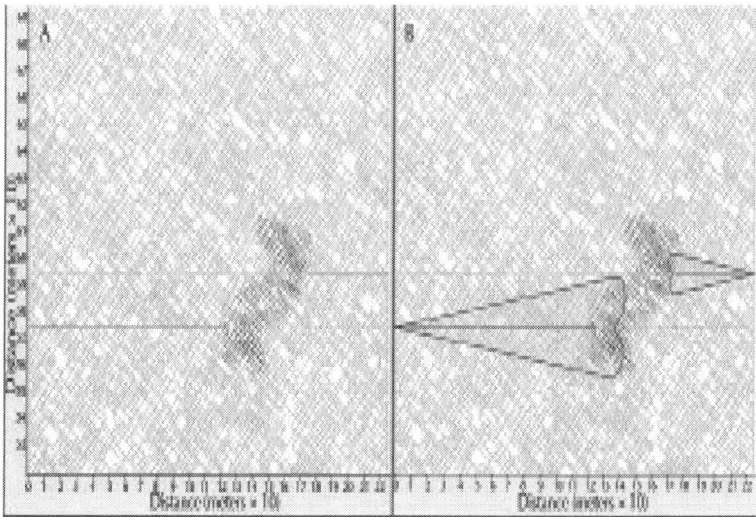

Figure 14: Natural fracture shear in blue from propagating hydraulic fractures Xf1 (from the left at 125m) and Xf2 (from the right at 50m) for a natural fracture friction of 35°.

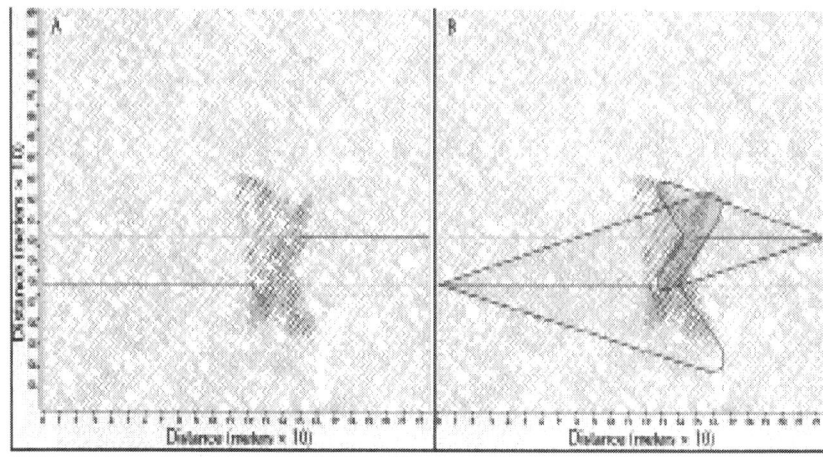

Figure 15: Natural fracture shear in blue from propagating hydraulic fractures Xf1 (from the left at 125m) and Xf2 (from the right at 75m) for a natural fracture friction of 25°.

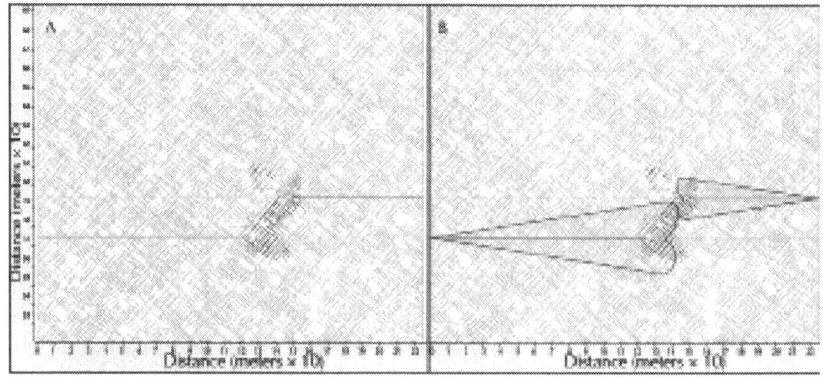

Figure 16: Natural fracture shear in blue from propagating hydraulic fractures Xf1 (from the left at 125m) and Xf2 (from the right at 75m) for a natural fracture friction of 35°.

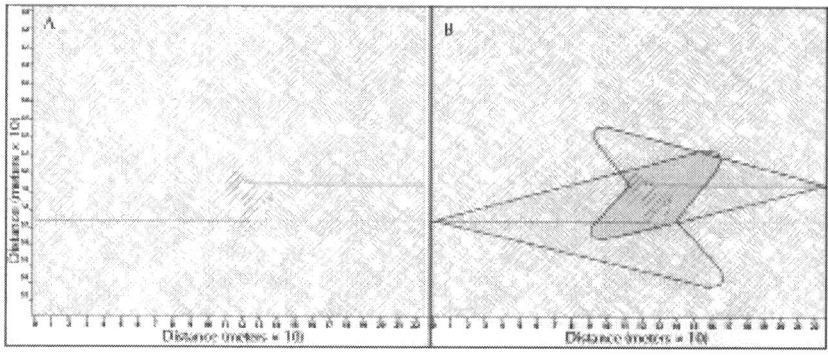

Figure 17: Natural fracture shear in blue from propagating hydraulic fractures Xf1 (from the left at 125m) and Xf2 (from the right at 100m) for a natural fracture friction of 25°.

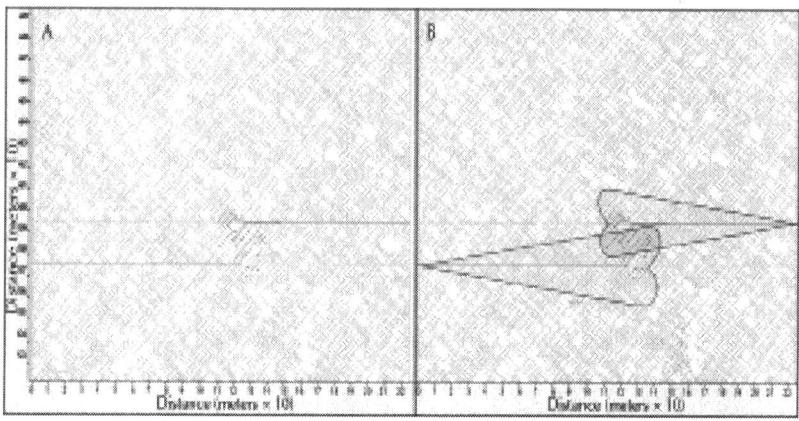

Figure 18: Natural fracture shear in blue from propagating hydraulic fractures Xf1 (from the left at 125m) and Xf2 (from the right at 100m) for a natural fracture friction of 35°.

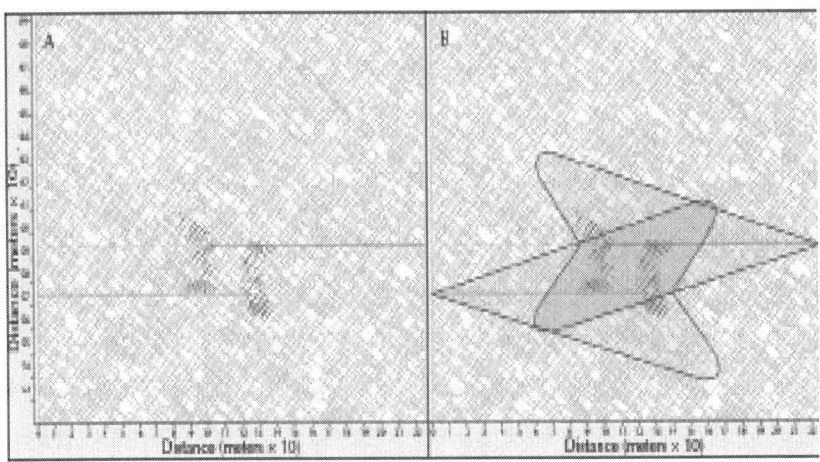

Figure 19: Natural fracture shear in blue from propagating hydraulic fractures Xf1 (from the left at 125m) and Xf2 (from the right at 125m) for a natural fracture friction of 25°.

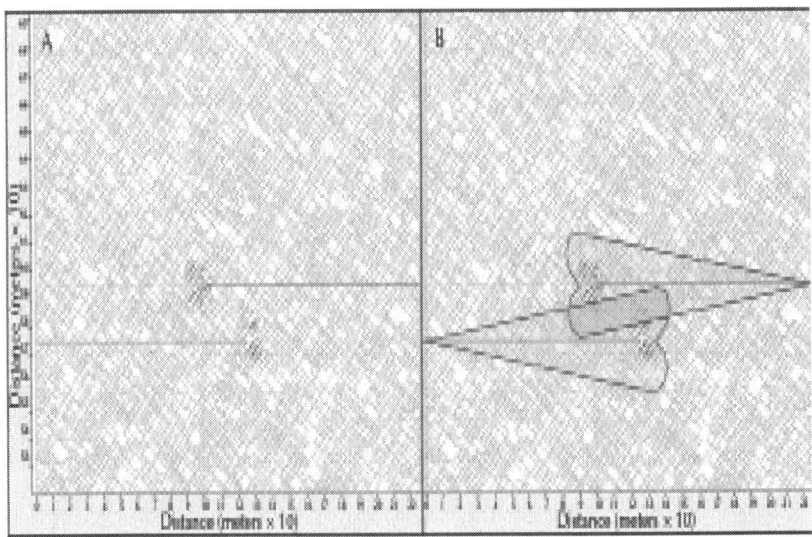

Figure 20: Natural fracture shear in blue from propagating hydraulic fractures Xf1 (from the left at 125m) and Xf2 (from the right at 125m) for a natural fracture friction of 35°.

Observations for the '145°' DFN and 20m Separation Simulations

The simulations were conducted such that hydraulic fracture Xf1 always had a fracture half-length of 125m and 'snapshots' were taken for hydraulic fracture Xf2 half-lengths of 50m, 75, 100m, and 125m. The simulated wellbores that Xf1 and Xf2 propagated from were set at 225m apart so that once Xf2 reached 100m or longer, it overlapped hydraulic fracture Xf1.

The significant observations from the simulation results include:

- For the 20m separation cases shown, the greatest 'extra' natural fracture shear (shear beyond what would have occurred from two independent hydraulic fractures) occurred when Xf2 was 50m in length. This was true for both natural fracture friction cases (Figures 13 and 14).

- As Xf2 grew beyond 50m in length, the 'extra' formation shear decreased and, most importantly, when Xf2 was 100m or 125m in length, there was a net loss of sheared natural fractures as compared to two independent hydraulic fractures.

- When Xf2 was 100m in length (so that the fracture tips from Xf1 and Xf2 just overlapped), the effect was a complete cancellation of natural fracture shear and a significant opening of natural fractures between Xf1 and Xf2 (likely allowing significant pressure communication) as shown in Figures 17 and 18.

- Once Xf2 exceeded 100m in length, natural fracture shear re-occurred, though it was significantly reduced (Figures 19 and 20). Note that in Figure 19 (natural fracture friction of 25°), the hydraulic fractures blunted the sheared fractures coming from the tip of the other hydraulic fracture acting as a form of release surface preventing transmission of shear on the other side of the hydraulic fracture.

Shear Results for the '145°' DFN and Other Hydraulic Fracture Separations

Figures 21 to 24 show a comparison of natural fracture shear for hydraulic fracture separations of 20m, 35m, and 45m for both natural fracture friction cases for Xf2 lengths of 75m and 125m.

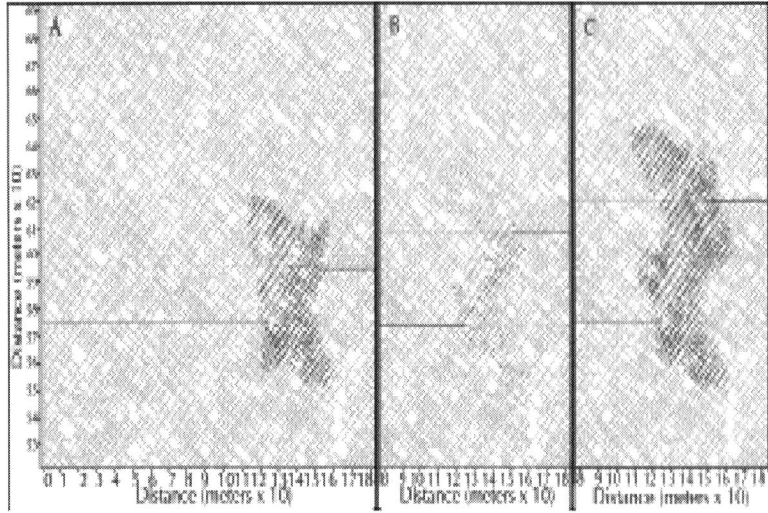

Figure 21: Natural fractures at shear as shown in blue for an Xf2 half-length of 75m and natural fracture friction of 25°. Red represents open fractures. A) A 20m hydraulic fracture separation; B) A 35m separation; and C) A 45m separation.

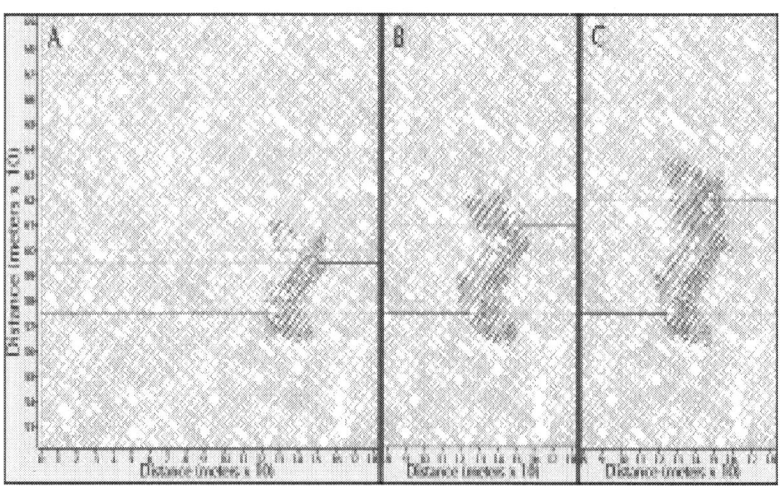

Figure 22: Natural fractures at shear as shown in blue for an Xf2 half-length of 75m and natural fracture friction of 35°. A) A 20m hydraulic fracture separation; B) A 35m separation; and C) A 45m separation.

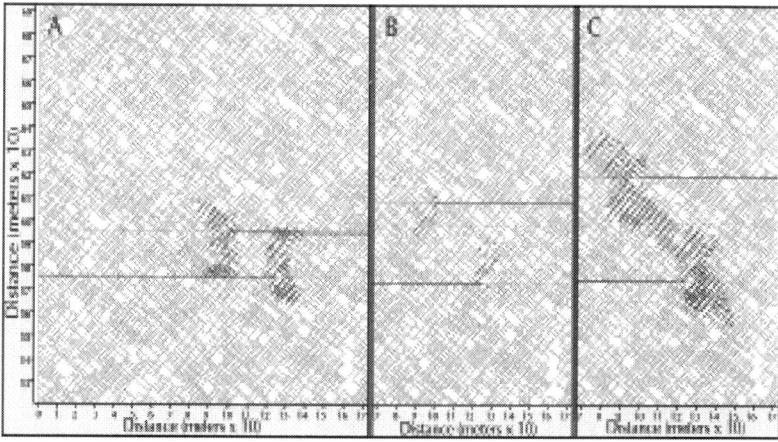

Figure 23: Natural fractures at shear as shown in blue for an Xf2 half-length of 125m and natural fracture friction of 25°. A) A 20m hydraulic fracture separation; B) A 35m separation; and C) A 45m separation.

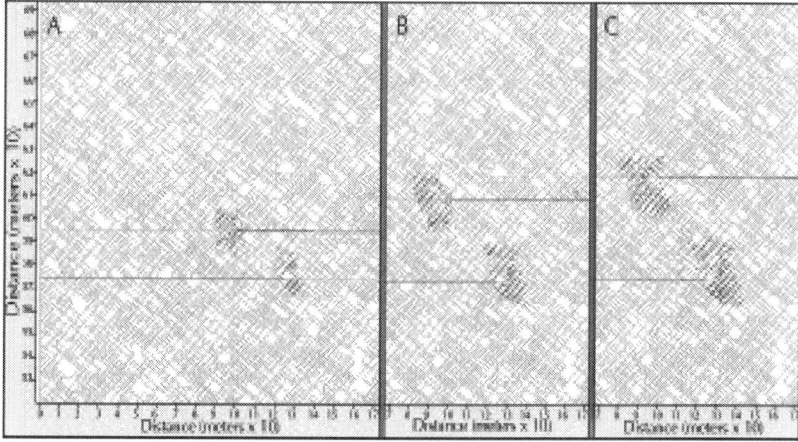

Figure 24: Natural fractures at shear as shown in blue for an Xf2 half-length of 125m and natural fracture friction of 35°. A) A 20m hydraulic fracture separation; B) A 35m separation; and C) A 45m separation.

Figure 25 presents a graph of the cumulative length of natural fracture shear for the 30 simulations with the '145°' DFN.

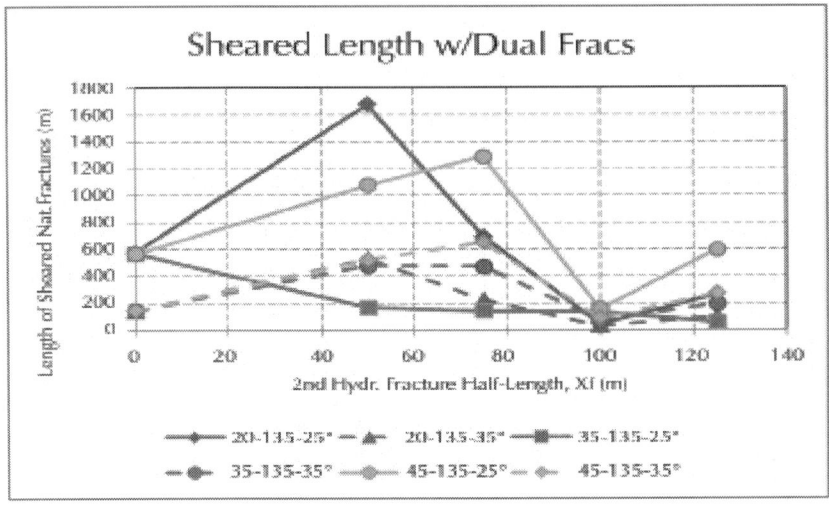

Figure 25: Graph of cumulative natural fracture shear length versus hydraulic fracture Xf2 half-length for separation cases 20m, 30m, and 45m for natural fracture friction of 25° and 35° for the '145°' DFN.

Observations for the '145°' DFN Dual Hydraulic Fracture Simulations

Within Figures 21 to 24, the simulation results for each of the three hydraulic fracture separation distances (20m, 35m, and 45m) are shown. Again, blue lines represent natural fractures at a shear condition at the moment the two hydraulic fractures are at their given half-length (125m for Xf1 and 75m or 125m for Xf2). Red lines represent open fractures (meaning there is no longer contact between the two sides of the fracture).

The significant observations from the simulation results include:

- Perhaps not surprisingly, the greatest total length of shear occurs for the 20m separation distance (at an Xf2 half-length of 50m); however, most interesting is that the total length of shear for the 45m separation distance is greater than that for the 35m separation distance. This suggests that natural fracture shear created between two hydraulic fracture tips is both a function of the separation distance and the orientation of the natural fractures.

- The simulation results suggest that the Xf2 half-length at which the maximum induced length of natural fracture shear occurs is related to the hydraulic fracture separation. For the 20m separation case, maximum shear occurred at Xf2 equal to 50m while for the 45m separation case, maximum shear occurred when the half-length of Xf2 was 75m.

- In all the cases, when the half-length of Xf2 was equal to 100m (so that the tips of Xf1 and Xf2 just overlapped), natural fracture shear was at a minimum.

- In all the cases, when the half-length of Xf2 grew to 125m, the cumulative length of natural fracture shear increased, but only modestly and significantly less than before the two hydraulic fractures overlapped. This suggests that overlapping hydraulic fractures do not enhance natural fracture shear but cause a net loss of shear relative to two independent hydraulic fractures.

- While for the 20m and 45m separation cases the effect of higher natural fracture friction was to significantly reduce the cumulative length of natural fracture shear (by 50% to 75%), for the 35m separation case the higher natural fracture friction resulted in greater cumulative natural fracture shear than the lower natural fracture friction case. While the full cause of this is not defined, a likely contributor is the ability of the rock mass in the low friction case to accommodate greater deformation without reaching the shear condition.

Shear Results for the '180°' DFN and 20m Hydraulic Fracture Separation

Figures 26 and 27 show the natural fracture shear (in blue) for Xf2 half-length cases of 50m, 75, 100m, and 125m for the '180°' DFN with a 20m separation distance and a natural fracture friction of 15°.

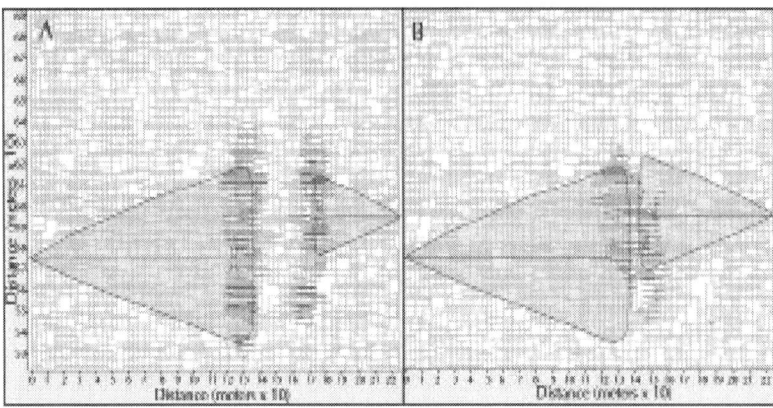

Figure 26: Natural fracture shear in blue from propagating hydraulic fractures Xf1 (from the left at 125m) and Xf2 (from the right) for a natural fracture friction of 15° and 20m hydraulic fracture separation. Red represents open fractures and shaded regions represent the expected shear area for two independent hydraulic fractures. A) Xf2 length equal to 50m; and B) Xf2 length equal to 75m.

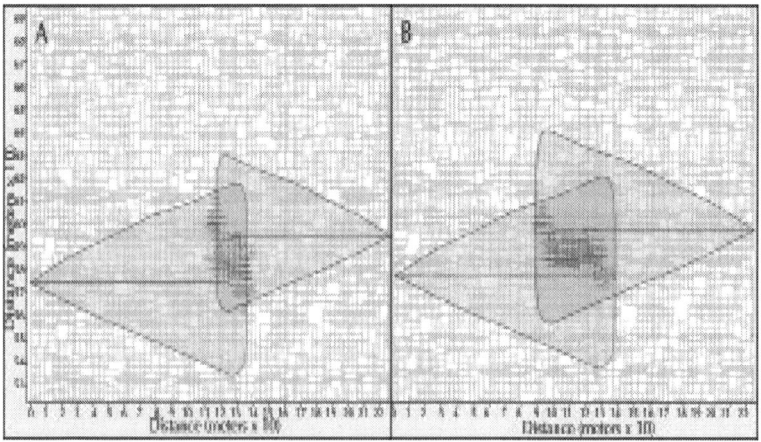

Figure 27: Natural fracture shear in blue from propagating hydraulic fractures Xf1 (from the left at 125m) and Xf2 (from the right) for a natural fracture friction of 15° and 20m hydraulic fracture separation. Red represents open fractures and shaded regions represent the expected shear area for two independent hydraulic fractures. A) Xf2 length equal to 100m; and B) Xf2 length equal to 125m.

Similar in fashion to Figures 13 to 20 for the '145°' DFN, Figures 26 and 27 show that there is an increase in natural fracture shear over two independent hydraulic fractures when Xf2 is less than about 75m. Further, when Xf2 exceeds a half-length of more than 75m (or, better, when the tip of Xf2 is within 25m of overlapping the tip of Xf1), then the result is a net loss of natural fracture shear over two independent hydraulic fractures.

Figures 28 to 31 show a comparison of natural fracture shear for hydraulic fracture separations of 20m, 35m, and 45m for both natural fracture friction cases (15° and 25°) for Xf2 lengths of 75m and 125m. Figure 32 shows a graph of the cumulative length of natural fracture shear versus Xf2 half-length for the 15° and 25° simulations (30 in total) for the '180°' DFN.

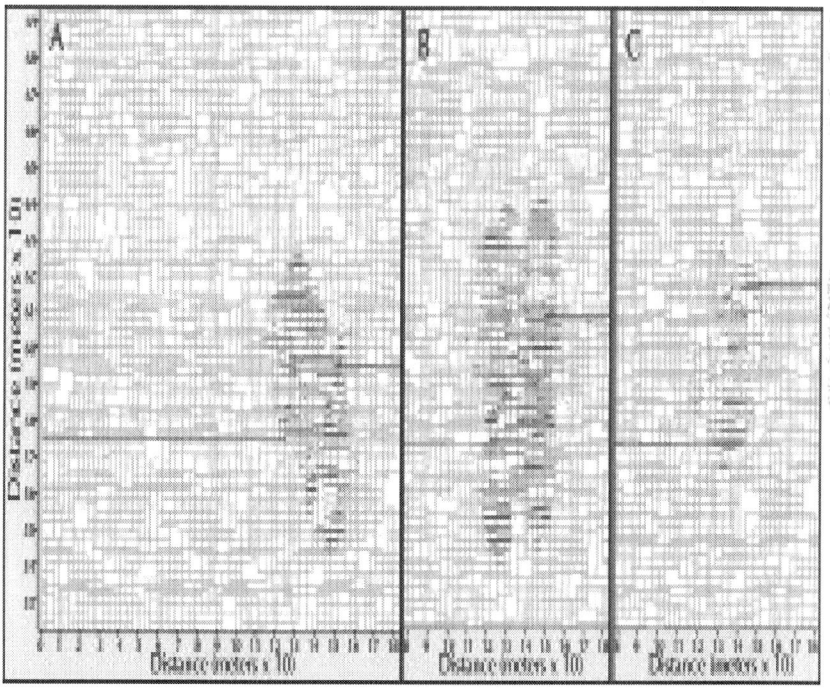

Figure 28: Natural fractures at shear (blue) for an Xf2 half-length of 75m and natural fracture friction of 15° for the '180°' DFN. Red represents open fractures. A) A 20m hydraulic fracture separation; B) A 35m separation; and C) A 45m separation.

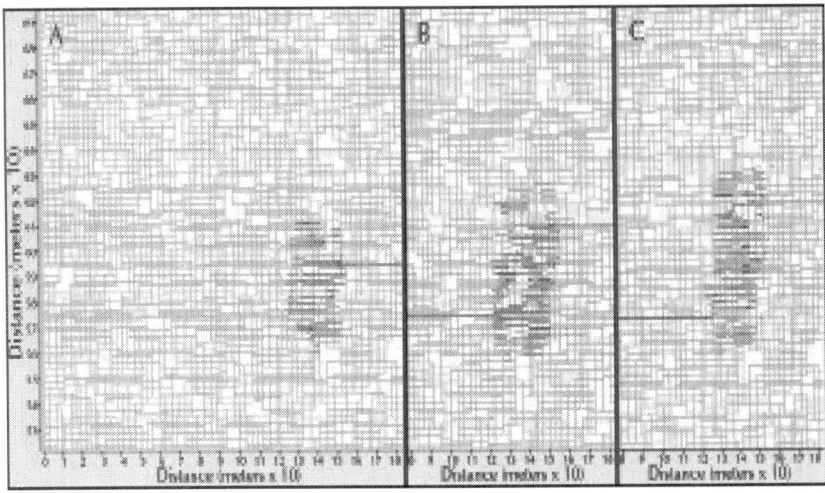

Figure 29: Natural fractures at shear (blue) for an Xf2 half-length of 75m and natural fracture friction of 25° for the '180°' DFN. Red represents open fractures. A) A 20m hydraulic fracture separation; B) A 35m separation; and C) A 45m separation.

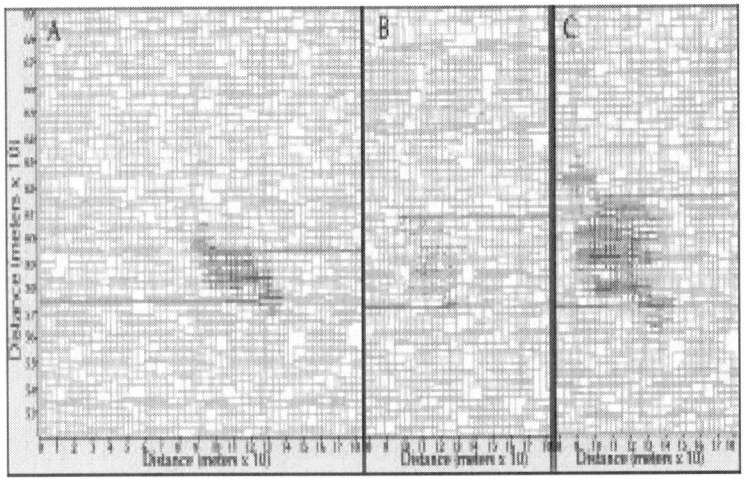

Figure 30: Natural fractures at shear (blue) for an Xf2 half-length of 125m and natural fracture friction of 15° for the '180°' DFN. Red represents open fractures. A) A 20m hydraulic fracture separation; B) A 35m separation; and C) A 45m separation.

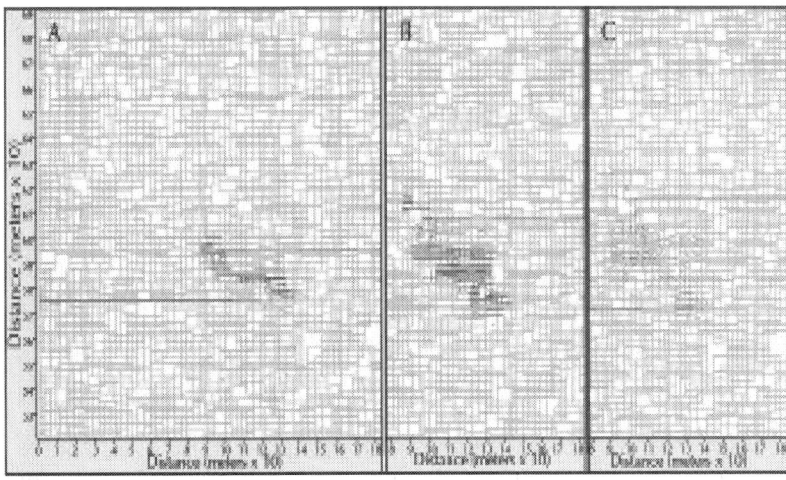

Figure 31: Natural fractures at shear (blue) for an Xf2 half-length of 125m and natural fracture friction of 25° for the '180°' DFN. Red represents open fractures. A) A 20m hydraulic fracture separation; B) A 35m separation; and C) A 45m separation.

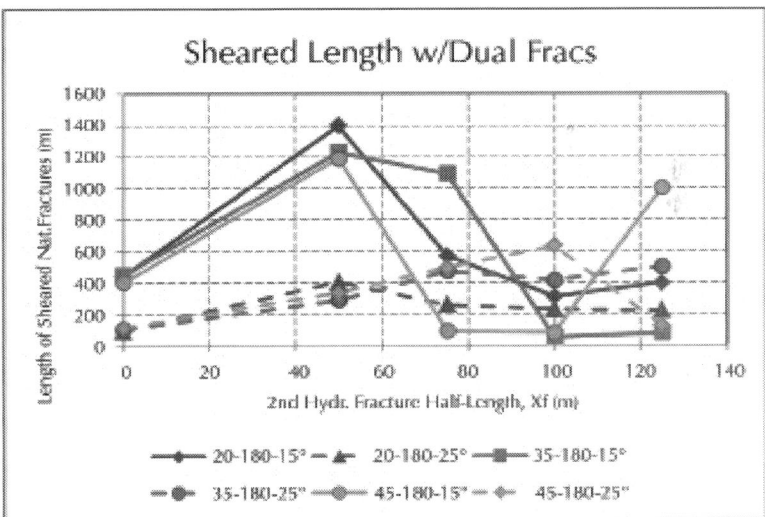

Figure 32: Graph of cumulative natural fracture shear length versus hydraulic fracture Xf2 half-length for separation cases 20m, 30m, and 45m for natural fracture friction of 15° and 25° for the '180°' DFN.

Observations for the '180°' DFN Dual Hydraulic Fracture Simulations

Within Figures 28 to 31, the simulation results for each of the three hydraulic fracture separation distances (20m, 35m, and 45m) are shown for the '180°' DFN. Again, blue lines represent natural fractures at a shear condition at the moment the two hydraulic fractures are at their given half-length (125m for Xf1 and 75m or 125m for Xf2). Red lines represent open fractures (meaning there is no longer contact between the two sides of the fracture).

The significant observations from the simulation results include:

- As with the simulations for the '145°' DFN, the higher friction cases generally resulted in less total length of sheared natural fractures than the lower friction cases; however, when Xf2 was 100m (so the tips of Xf1 and Xf2 just overlapped), the higher friction cases generally had more sheared length of natural fractures.

- For all three separation cases, the greatest total length of natural fracture shear occurred when the Xf2 half-length was 50m. As the separation distance increased between the hydraulic fractures, the total length of natural fracture shear became increasing sensitive to Xf2 half-length. For the 45m separation case, the total length of natural fracture shear decreased by more than 90% between an Xf2 half-length of 50m and 75m.

- For the '180°' DFN, the cumulative length of natural fracture shear was not as sensitive at an Xf2 half-length of 100m as was the '145°' DFN. This, again, shows that the orientation of the natural fractures is important in creating natural fracture shear when two hydraulic fractures influence each other.

- As with the '145°' DFN, the amount of open fractures in the '180°' DFN cases appeared to influence the amount of natural fracture shear. Further, open natural fractures will be more conductive and, likely, allow pressure communication between hydraulic fractures potentially causing screenout events.

Shear Results for the '180°' DFN and Altered in-Situ Stress

Recall from Table 2 that a number of simulations were conducted with the '180°' DFN model wherein the in-situ stress field was altered. As shown in Table 2, the vertical stress Sv, maximum horizontal stress SHmax, and pore pressure were kept constant and the minimum horizontal stress was increased by 5.6 MPa, which resulted in near-isotropic horizontal stress conditions. Figures 33 and 34 show the sheared natural fractures for the 20m separation case and natural fracture friction of 15° and 25° and with the revise in-situ stress. Figure 35 shows a graph of the length of natural fracture shear for the 25° simulations from and Table 6 and initial and revised stresses.

Observations for the '180°' DFN Dual Fracture Simulations with Revised in-Situ Stress

The significant observations from the simulation of the change in in-situ stress include:

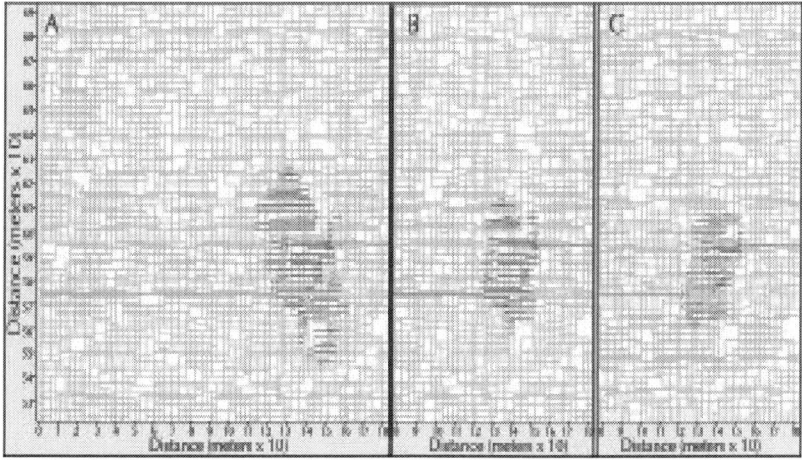

Figure 33: Natural fractures at shear (blue) for an Xf2 half-length of 75m for the '180°' DFN at a 20m separation. Red represents open fractures. A) 15° natural fracture friction; B) 25° fracture friction; and C) 25° fracture friction and revised in-situ stress.

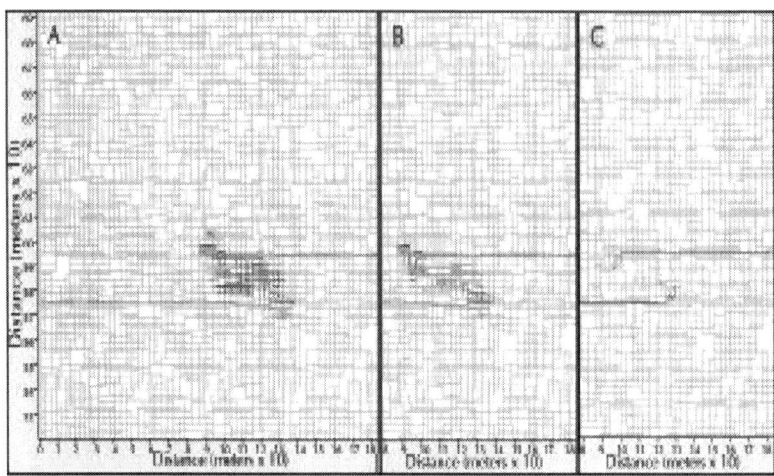

Figure 34: Natural fractures at shear (blue) for an Xf2 half-length of 125m for the '180°' DFN at a 20m separation. Red represents open fractures. A) 15° natural fracture friction; B) 25° fracture friction; and C) 25° fracture friction and revised in-situ stress.

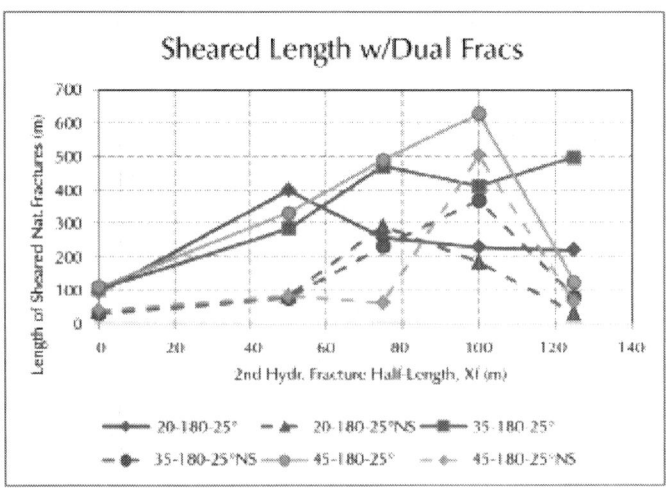

Figure 35: Graph of natural fracture shear length versus hydraulic fracture Xf2 half-length for separation cases 20m, 30m, and 45m for natural fracture friction of 25° for the '180°' DFN with the initial and revised in-situ stress (see Table 2).

- As shown in Figure 35, the maximum cumulative length of natural fracture shear occurred at the 45m separation distance for either in-situ stress case.

- Clearly, moving towards more of an isotropic in-situ horizontal stress reduced the total length of natural fracture shear. Furthermore, the overall behavior also changed such that the maximum cumulative length of natural fracture shear occurred at 100m (the point of tip-to-tip overlap) for the near-isotropic stress case.

- Particularly at larger separation distances (the 45m separation case), the isotropic in-situ stress appeared to make the two hydraulic fractures cancel the shear from each other until the tips of Xf1 and Xf2 were close or overlapped.

- Perhaps even more so than the initial stress cases, the cumulative length of natural fracture shear when the tips of Xf1 and Xf2 overlapped (e.g., the 125m Xf2 case) dropped to near-zero for the revised stress simulations. This suggests that, even with a revised in-situ stress field, overlapping the tips of hydraulic fractures from parallel wellbores creates a net loss of natural fracture shear.

DISCUSSION

The goal of the effort was to quantitatively consider the change in natural fracture shear (shear being analogous with microseismicity generation and the potential stimulation of the natural fractures providing increased production from the hydraulic fracture) for multi-well completions. It is commonly believed that by configuring the geometry and injection behavior from parallel wellbores (e.g., simultaneous fracturing, zipper-fracs, and modified zipper-fracs), shear of the natural fractures can be enhanced (thereby increasing production).

During the evaluations presented in this paper, the following parameter effects were considered:

- Fracture network orientation (i.e., the '180°' DFN and the '145°' DFN);

- Natural fracture friction angle (15° and 25° for the '180°' DFN and 25° and 35° for the '145°' DFN);

- Hydraulic fracture separation (offset between injection points) from 20m to 45m;
- Hydraulic fracture half-length from the second wellbore (Xf2 helf-lengths of 50m to 125m); and
- In-situ stress (from a horizontal stress ratio – SHmax/Shmin - of 1.18 to a ratio of 1.03).

Observations on the Influence of Fracture Network

As shown in Figures 6 and 8, the natural fracture shear pattern coming off the tip of a propagating hydraulic fracture depends upon the orientation and nature of the natural fracture system. For the '180°' DFN, natural fracture shear extended a bit beyond the hydraulic fracture tip, but mainly lay in a symmetrical pattern perpendicular to the direction of hydraulic fracture propagation. In contrast, for the '145°' DFN, the natural fracture shear pattern was asymmetric and lead the tip of the propagating hydraulic fracture. Clearly as observed in previous publications (Nagel et al. 2011a), interpreting microseismic event locations cannot be done without consideration of the general orientation of the natural fracture pattern.

The natural fracture pattern also plays a role in the amount of natural fracture shear (and, by analogy, the number of microseismic events). For the same natural fracture friction (and same in-situ stress), the total area of natural fracture shear for the '180°' DFN was only 42% of that for the '145°' DFN (2220m^2 versus 5250m^2). However, as shown in the graphs in Figures 25 and 32, the overall trends in the cumulative length of natural fracture shear from dual hydraulic fractures was similar (with the exception of the 35m spacing for the '145°' DFN in which natural fracture shear was very low).

Observations on the Influence of Natural Fracture Friction

As evident from the figures of natural fracture shear and the quantitative results in Figures 25 and 32, natural fracture friction plays a significant role in determining the amount of natural fracture shear (and, by analogy, the number of microseismic events). The influence of natural fracture friction also depends upon the underlying natural fracture pattern (and stress ratio). The area of shear generated for the '180°' DFN at a friction angle of 15° was nearly equal to the area of natural fracture shear for the '145°' DFN at a friction angle of 25° (5740m^2 versus 5250m^2).

Less so for the '145°' DFN and more so for the '180°' DFN, the higher fracture friction tended to push the point of maximum total length of natural fracture shear towards longer Xf2 half-lengths; however, these longer half-lengths also represented conditions when there was a net loss of natural fracture shear for two dual hydraulic fractures over two equivalent independent hydraulic fractures.

Observations on the Influence of Hydraulic Fracture Separation Distance

Figures 25 and 32 suggest that hydraulic fracture separation did not significantly affect the maximum total length of natural fracture shear (more so for the '180°' DFN and less so for the '145°' DFN). However, the influence of separation spacing was more apparent when the Xf2 half-length was 100m or longer.

Though Figures 25 and 32 may suggest a somewhat limited influence of hydraulic fracture spacing, this is clearly not the whole picture. As shown in Figures 9 through 12 in particular, and somewhat in Figures 13 through 20, the critical issues for hydraulic fracture separation are to: 1) shear as much total formation as possible; and 2) not cause a net loss of natural fracture shear by placing hydraulic fractures too close. Figure 10 shows that at a 45m hydraulic fracture separation distance (for dual, 125m-long hydraulic fractures and a natural fracture friction angle of 15°) the shear area from the two hydraulic fractures still overlapped (when the hydraulic fractures act independently). In contrast, Figure 12 shows that a 45m separation distance may be too much when natural fracture friction angle is 25°.

Observations on the Influence of Hydraulic Fracture Xf2 Half-Length

The simulation results (especially Figures 25 and 32) show that the amount of natural fracture shear is significantly influenced by the half-length of the Xf2 hydraulic fracture in a dual fracture configuration. The overall trend of the results is that keeping the half-length of Xf2 small enough to prevent the tip of Xf2 from getting closer than 25m to the tip of Xf1 (that is, no overlap of the hydraulic fractures) creates the maximum total length of natural fracture shear. Further, as shown inFigures 9 through 12, keeping the half-length of Xf2 small enough may also cause a net increase in natural fracture shear (over that from two independent hydraulic fractures), which is the goal of a dual hydraulic fracture configuration.

CONCLUSIONS

- As natural fracture orientation (relative to the orientation of a hydraulic fracture) significantly influences the amount and location of natural fracture shear, multi-well completion optimization (wherein the goal is to maximize natural fracture shear, i.e., maximize 'complexity') requires the evaluation and consideration of natural fracture orientation.

- As natural fracture friction controls the depth and amount of natural fracture shear, multi-well completion optimization requires the evaluation and consideration of natural fracture friction properties.

- The optimum hydraulic fracture separation distance for multi-well completions (i.e, the separation of hydraulic fractures along their respective wellbores) must be determined in consideration of natural fracture properties (e.g., orientation and friction properties) and the in-situ stress ratio.

- For multi-well completion schemes, the design length of the second hydraulic fracture (Xf2) should be kept less than the point of overlap with the first hydraulic fracture (Xf1) and be optimized in conjunction with the hydraulic fracture separation distance.

- Overall, the simulation results presented suggest that there is the potential for only modest improvements in stimulation complexity from the modified zipper-frac completion scheme while the potential for well-to-well communication (and possible screenout conditions) increases.

REFERENCES

1. K Agarwal, M. J Mayerhofer, and N. R Warpinski, 2012Impact of Geomechanics on Microseismicity", Paper SPE 152835 presented at the SPE/EAGE European Unconventional Resources Conference and Exhibition, Vienna, Austria, 2022March.

2. Clover Global SolutionsLP, 2012The Seven Major U.S. Shale Plays", http://c1wsolutions.wordpress.com/2012/09/13/the-seven-major-u-s-shale-plays

3. GroundWaterProtectionCouncilandALLConsulting2009Modern Shale Gas Development in the United States: A Primer", prepared for the US DOE, Office of Fossil Energy, DE-FG2604NT15444.

4. G. E King, 2010Thirty Years of Gas Shale Fracturing: What Have We Learned?", Paper SPE 133456 presented at the SPE Annual Technical Conference and Exhibition, Florence, Italy, 1922September.

5. N Nagel, and M Sanchez-nagel, 2011Stress Shadowing and Microseismic Events: A Numerical Evaluation", Paper SPE 147363 presented at the SPE Annual Technical Conference and Exhibition, Denver, Colorado, USA, 30 October-2 November.

6. N Nagel, B Damjanac, X Garcia, and M Sanchez-nagel, 2011bDiscrete Element Hydraulic Fracture Modeling- Evaluating Changes in Natural Fracture Aperture and Transmissivity", Paper SPE 148957 presented at the Canadian Unconventional Resources Conference, Calgary, Alberta, Canada, 1517November.

7. N Nagel, I Gil, M Sanchez-nagel, and B Damjanac, 2011aSimulating Hydraulic Fracturing in Real Fractured Rock-Overcoming the Limits of Pseudo3D Models", Paper SPE 140480 presented at the SPE Hydraulic Fracturing Technology Conference and Exhibition, The Woodlands, Texas, USA, 2426January.

8. N Nagel, M Sanchez-nagel, and B. T Lee, 2012aGas Shale Hydraulic Fracturing: A Numerical Evaluation of the Effect of Geomechanical Parameters", Paper SPE 152192 presented at the SPE Hydraulic Fracturing Technology Conference and Exhibition, The Woodlands, Texas, USA, 68February.

9. N. B Nagel, X Garcia, B Lee, and M Sanchez-nagel, 2012dHydraulic Fracturing Optimization for Unconventional Reservoirs- The Critical Role of the Mechanical Properties of the Natural Fracture Network", Paper SPE 161934 presented at the SPE Canadian Unconventional Resources Conference, Calgary, Alberta, Canada, 30 October- 1 November.

10. N. B Nagel, M Sanchez-nagel, F Zhang, X Garcia, and B Lee, 2013Coupled Numerical Evaluations of the Geomechanical Interactions Between a Hydrualic Fracture Stimulation and a Natural Fracture System in Shale Formations", Rock Mechanics and Rock Engineering, pending publication.

11. N. B Nagel, M Sanchez-nagel, X Garcia, and B Lee, 2012bA Numerical Evaluation of the Geomechanical Interactions Between a Hydraulic Fracture Stimulation and a Natural Fracture System", ARMA 12287presented at the 46th Rock Mechanics / Geomechanics Symposium, Chicago, Illinois, 24-27 June.

12. N. B Nagel, M Sanchez-nagel, X Garcia, and B Lee, 2012c Understanding, SRV": A Numerical Investigation of "Wet" vs. "Dry" Microseismicity During Hydraulic Fracturing", Paper SPE 159791 presented the SPE Annual Technical Conference and Exhibition held in San Antonio, Texas, USA, 81OOctober.

13. I. N Sneddon, 1946The Distribution of Stress in the Neighbourhood of a Crack in an Elastic Solid", Proc. R. Soc. London, Ser. A. 195229260

14. U.S. Energy Information Administration, 2012, "Annual Energy Outlook 2012 Early Release Overview", U.S. Dept. of Energy, Washington D.C., USA, www.eia.gov

15. U.S. Energy Information Administration, 2013, "Annual Energy Outlook 2013 Early Release Overview", U.S. Dept. of Energy, Washington D.C., USA, www.eia.gov

Citations

CHAPTER 1

John Wei-Shan Hu, Yi-Chung Hu, and Chien-Yu Lin, "Effect of Temperature Shock and Inventory Surprises on Natural Gas and Heating Oil Futures Returns," The Scientific World Journal, vol. 2014, Article ID 457636, 10 pages, 2014, doi:10.1155/2014/457636.

CHAPTER 2

Barry Goldstein, Michael Malavazos, Alexandra Wickham, Michael Jarosz, Dominic Pepicelli, Mieka Webb and Dale Wenham (2013). Regulatory Nirvana for Hydraulic Fracture Stimulation, Effective and Sustainable Hydraulic Fracturing, Dr. Rob Jeffrey (Ed.), ISBN: 978-953-51-1137-5, InTech, DOI: 10.5772/56381.

CHAPTER 3

Xiaochun Zhang, Nathan P Myhrvold, and Ken Caldeira, Key Factors for Assessing Climate Benefits of Natural Gas Versus Coal Electricity Generation, Doi:10.1088/1748-9326/9/11/114022.

CHAPTER 4

Hartmut Wendtl, Estevam V. Spinacél, Almir Oliveira NetoII, and Marcelo Linardi, Electrocatalysis and electrocatalysts for low temperature fuel cells: fundamentals, state of the art, research and development, http://dx.doi.org/10.1590/S0100-40422005000600023.

CHAPTER 5

André Vagner Gaathaug, Dag Bjerketvedt, Knut Vaagsaether, and Sandra Hennie Nilsen, "Experimental Study of Gas Explosions in Hydrogen Sulfide-Natural Gas-Air Mixtures," Journal of Combustion, vol. 2014, Article ID 905893, 12 pages, 2014. doi:10.1155/2014/905893.

CHAPTER 6

Walter Holweger (2013). Fundamentals of Lubricants and Lubrication, Tribology - Fundamentals and Advancements, Dr. Jürgen Gegner (Ed.), ISBN: 978-953-51-1135-1, InTech, DOI: 10.5772/55731.

CHAPTER 7

N. Nagel, F. Zhang, M. Sanchez-Nagel and B. Lee (2013). Quantitative Evaluation of Completion Techniques on Influencing Shale Fracture 'Complexity', Effective and Sustainable Hydraulic Fracturing, Dr. Rob Jeffrey (Ed.), ISBN: 978-953-51-1137-5, InTech, DOI: 10.5772/56304.

Index